THREE WAYS TO VIEW THE WORLD

Way Outside the Rigid Box of Ordinary Thinking
about
Our Human Strivings on Earth Today and Evermore

WILLIAM SALO

An archival organization of a selected group of thoughtful writings and an attempt to understand the human role in the universe

BALBOA.
PRESS

A DIVISION OF HAY HOUSE

Balboa Press books may be ordered through booksellers or by contacting:

Balboa Press
A Division of Hay House
1663 Liberty Drive
Bloomington, IN 47403
www.balboapress.com
1-(877) 407-4847

ISBN: 978-1-4525-5730-4 (sc)
ISBN: 978-1-4525-5732-8 (hc)
ISBN: 978-1-4525-5731-1 (e)

Library of Congress Control Number: 2012915226

Because of the dynamic nature of the Internet, any web addresses or links contained in this book may have changed since publication and may no longer be valid. The views expressed in this work are solely those of the author and do not necessarily reflect the views of the publisher, and the publisher hereby disclaims any responsibility for them.

The author of this book does not dispense medical advice or prescribe the use of any technique as a form of treatment for physical, emotional, or medical problems without the advice of a physician, either directly or indirectly. The intent of the author is only to offer information of a general nature to help you in your quest for emotional and spiritual well-being. In the event you use any of the information in this book for yourself, which is your constitutional right, the author and the publisher assume no responsibility for your actions.

Any people depicted in stock imagery provided by Thinkstock are models, and such images are being used for illustrative purposes only.
Certain stock imagery © Thinkstock.

Printed in the United States of America

Balboa Press rev. date: 09/10/2012

Twixt commonplace
And state of grace
Lie a million miles of
Hallowed space.
Tis where all humans,
Here and there,
Must challenge life
To care and share.
And those who have a
Thinking mind
Forge on and leave
Numbskulls behind.

Three Ways to View the World for All Thinkers, Seniors, and Cosmic Pioneers

Sector Captions

MACROSCOPIC TOPICS . . . toeholds up the matterhorn of personal maturation while humans seek for Fame and Fortune in their lives.

Macroscopic Topics portray the challenges of climbing the mountain of life from infancy, through adolescence, into adulthood and to the productive years of heavy lifting necessary to reach the acme of success, perhaps fame and fortune, at the mountaintop. The condition of the world, its commerce and politics, and the ratio of educated, intellectual thinkers influencing the erudition of the masses has grown and is changing dramatically almost day by day. It is affecting communications technology and access to a cosmic way of analysis that is promising an exciting and productive future for both young and old. It takes courage, perseverance, judgement and ambition to scale the mountain to whatever altitude any human's aspirations can perceive.

EXX-RAYS FROM THE AGING CAGE . . . coping with the declivity of Super Seniordom.

Exx-Rays from the Aging Cage addresses the trials in growing older but also details the opportunities and pleasures to be found in the senior years of life's time clock. The philosophical color has been derived from the experiences of those who are super seniors, that is, those who have and are still coping with the vicissitudes of life on this ball of mud we call planet Earth. Most are or have occcupied space and time here on Earth for over eighty years or more, providing a veritable kaleidoscope of real experiences.Veterans of World War II have formed associations to share past experiences and to prize present camaraderie. The experiences of shifting from a controlled environment to an untested challenge of personal management calls for a sea change in mental and physical attitudes. Retirement years can be uplifting with a variety of experiences that were too hard to effect during the earning years. The Aging Cage can be very confining in one state of mind, but also quite open and exhilarating in another.

COSMIC CYBERTREKS . . . subjective exploration in the magna void of Cosmic space.

Cosmic Cyber Treks is intended to be mind-boggling and will stretch imaginations into new areas quite outside the box of normal comfort zones in traditional beliefs and understanding. The messages are not intended to proselytize or be tutorial, but rather wish to introduce new ways of thinking about the universe, the world, and the place in it for Homo sapiens. The pressure of change is central to this examination. The possibility of new thinking on the structure and purpose of the universe and how such different ideas might affect the cosmic impact on planet Earth are basic in this proffering of existential thinking. The movement of human thought has been changed drastically via the impetus of the power in electronic communications technology. The movement away from reading printed material for information has been taken over by television, social media, interactive electronic music, and the interpersonal networks of telephone and texted communications. With the more laborious technique of reading books and other literature that demand concentration of thought and mental word pictures to be understood, the process of communications has evolved to be entertaining observers rather than motivating them to think. That change has claimed the interest of the younger generations who place a superior value on fun at the expense of more serious cogitation. The human mind, like every other muscle in the human body, needs rigorous exercise to combat its denigration to where its possessors are doomed to become bystanders rather than players in future major evolutionary developments. With this sword of Damocles hanging over the head of all concerned members of the human cast, there is strong motivation to bring back the mental exercise of pure thinking. With the realization that this motivation cannot be dictated to, but must be suggested, the writings in this book are an attempt to test these principles. It is hoped that the essence, if not the concept, will load the weapons of thinking among humans in future years. The Universe does not reveal its secrets to those minds that try to rationalize that events in Cyber Space occur as planetary extensions rather than as just the other way around. Planetary experiences are products of Cosmic causes and therefore should portray the thought that the human mind must expand its capacity to encompass the scope of outer space if logic is to be the order of the day.

Macroscopic Topics Contents

Twixt commonplace
And state of grace
Lie a million miles of
Hallowed space.
Tis where all humans,
Here and there,
Must challenge life
To care and share.
And those who have a
Thinking mind
Forge on and leave
Numbskulls behind.

Reading Sources for General Information

The Warren Buffett Way	R.G. Hagstrom Jr.
The Fifth Discipline	P.M. Senge
The Strategic Mind	R. Gorzynski
The New Rules	J.P. Kotter
The Enneagram Advantage	H. Palmer and P.B. Brown
The Power of Now	E. Tolle
Tougher Boards for Tougher Times	W.A. Dimma

Mankind's Ego Need: The Mighty Muscle of Motivation

All true motivation in the *macro* book is deeply rooted in the visceral egos of all humankind on Earth. The ego, a demanding, schizophrenic bitch, is centered on two main characteristics; pride and vanity. To most humans, these are often thought to be the same. Not true.

To be well rounded, egotistically speaking, is not merely necessary but is mandatory for all motivated individuals in Earth's highly competitive, widely varied personal environments. However, the weasel in the woodpile that sits in judgment calls for each human to find equitable balance between pride and vanity, bringing balance to the decisions that all must make on our highways of destiny, so that the fates and fortunes in our lives can be positive and productive. The challenges are several and hold us fully responsible for what we do and become in our individual lives.

Maintaining a sense of pride is a tough and often lonely mindset, and requires self-instilled discipline. Mostly, the only reward is a measure of self-satisfaction that must be covert if it is to avoid a segue into vanity. Pride is experienced and sustained from a nucleus of successful conquests over challenges that have great importance to human existence. There is pleasure in such accomplishments and, hence, marked pride settles within the soul.

The gales of vanity take little effort to weather. One can enjoy the accolades by being openly receptive to the wash of popularity, real or devised, that can come sweeping through one's cognizance and into willing acceptance. Performances in events of real or imagined interest to idol worshippers begets a sense of self-worth that is overwhelming and addictive. If it is not harshly vetted to separate fact from fiction, common sense goes down the drain, and a very illusory feeling of regal importance takes over. Once bitten by the asp of vanity, a return to real values is only found in the cathartic scrutiny of criticism. Decamping from the pedestal of fame is never a life-forgiving exercise. The payoff is often depression, alcoholic nightmares, and a serious mental fall from grace. Now, one might think that vanity has no redeeming graces. That would be too harsh a criticism. For instance, all great performers on public stages need the feelings of importance that are intrinsic in the florid atmosphere of artists, politicians, and authors. They often must evince emotions that are not their own. If one is to reach the deepest wells of emotional talent, supporting credibility in a world of creative incredibility, then vanity is the plinth on which success can support a productive role.

To affect a meaningful role in business or artistic worlds, regular, personal evaluation of the balance between pride and vanity is mandatory. If crossover does occur, when the wrong virtue dominates unsuitable milieus, the price that is paid is a killer in due course. Many examples have laced the history of personal downfalls from such exercises. The human ego is a tough taskmaster when it comes to behavioral choices. If the choice between a private role of pride and the public role of vanity is wisely utilized, the end product of success in careers, and life itself, is well endowed. The alternative is what is called mega ego (ME). It evolves when its possessor craves godlike adulation from all and sundry. Most ME types immolate themselves or else lose their sanity.

Pride or vanity? To be or not to be, that is the question. It bears some cosmic thinking at a deeper level in the human mind. This contemplative task cannot be foisted on outsiders. It stays at home. And that is the rub. It is easier to revel in the praise, fair or phony, of vanity than to irk beneath the rasp of pride. It depends on how one sees the face in the mirror of one's own reality.

Readers Space for Thoughts and Ideas.

Cosmic Allusions Attracted by Subconscious Cybernetic Symbiosis

In the workaday world, where most of human laypersons dwell, the use and even the understanding of upper-crusty words are rarely experienced and, hence, are often resented. However, in those milieus where their use is daily dialogue, they are very special and succinct in what they do express.

In these scientific and philosophical labs, cybernetics has various connotations. In the field of abstract ideation, they swirl mostly around the definition of sensations and how these may relate to one another in the human mind. The application of principles and a way to communicate and control is at the heart of the defining word. Denotations of feelings, as opposed to facts, are at the core.

Symbiosis, as it is understood in erudite circles, defines the way in which two organic or inorganic entities relate and communicate with each other. This does not suggest that electronic systems, chemical components, human beings, and flora and fauna of all types do not have methods of reaching and coordinating with each other.

Humans are conditioned, animistic creatures, but being mainly pragmatic, the general feeling is that, if it is possible to see, hear, smell, touch, or taste, these components limn the human vision of reality. While fiduciary acceptance of possible life forces beyond the human mind and body might exist in the all-pervasive energy of the universe, it calls for masses of philosophical mental energy to believe outright that cosmic forces play a role in everyday human life.

But many must, as insight is gained into the apparent pathways of human destinies, be forced to realize just how intricate and complex the electronic system of the mind must be. In certain states of behavior, the mental processes seem able to be in contact with the cosmic energy that flows through the human systems every second of existence on Earth. Humans recognize that their minds have two separate but co-ordinated roles:

- The obvious condition is the conscious one that is known as the command center for all the functions that are required 24/7 over a lifetime. This role sets thoughts into focus, orders muscles into action, and monitors both pain and pleasure to the extent that all are led to believe these actions are the basis of all human-recognized reality.

- The second role is that of the subconscious. Its functions are programmed into a newborn's mind at birth. All the basic factors that are needed for human survival to create, use, and sustain the energies of life are automatically included. The nuances and essential tools of more subtle events are there to give emotional texture to living, thinking, creating, and remembering. The possibility is that this side of the mind does not act unilaterally but explores the inventories of the Universe for ideas and solutions from time to time. The scope of human imagination is extensive and dynamic and can visualize the effect that direct contact with this cosmic energy might have on one's daily existence. Perhaps it is without human awareness or critical control over related events.

To possess some logic demands deep, or "out of the box," thinking. From the perspective of a professional in the field or other interested seekers, experience with unexplained mental events at critical moments encourages some examination. These can occur when mulling over an intransigent refusal of the conscious mind to supply a solution to a sticky problem or issue. After some frustrated effort, with deep focus to define the crux, and a wish for a probable solution, the event is put aside to rest in mental limbo. In due course, much to the surprise and delight of the proposer, the answer pops up in the conscious mind, unrequested. Though unaware of the process, this is a typical example of cosmic allusion. In fact, the contact of the sub-conscious mind with the electromagnetic inventory of knowledge in the magna vault of the Universe through cybernetic symbiosis, created the desired response, complete with the solution.

Energy of all types, once created, can never be destroyed. It must be utilized in alternate forms or returned to the cosmic inventory of the Universe in one form or another. Humans, when they arrive as infants, absorb an amount of this energy to support the rigors of survival and will continue to develop new energy as they progress on their treks of destiny as adults. This energy must then return to the Universe when the human body expires. This is thought to be in the form of electromagnetic particles that are used and reused, ad infinitum, by the cosmic processes. This concept can be controversial because of alternative, long-term beliefs.

It would seem that the human brain cannot demand cosmic allusion. It works independently from human commands but is covertly supplied through the subconscious network only when the need is focused and its fulfillment is left up to the cosmos to effect. Any doubting Thomas will not be able to reach this pool of energy. The autocrat will miss the boat. To gain entry to the inventory in the magna vault of the Universe, calls for intense focus on the issue or problem but not on the proposed solution. That part will be exercised in cosmic space that, as yet, is not a free, open door to fulfill humans' commands.

It can be speculated that the huge fund of growing knowledge in the Universe will become more freely available to humankind when Homo sapiens start to believe in the possibility, learn the methods of knowledge retrieval, and mature sufficiently to honor the principles of sharing cosmic experience with all. Greed and inconsideration will not find the key to the cosmic energy vault. The Cloud system of storing digital information in space, a very new form of electronic wizardry, may be a first strike in the building of the needed bridge.

Cosmic allusion has been known to act as a guardian of human life in situations of dire peril. Evaluation of risk and a warning of consequences at decisive, critical moments have been experienced as an undeniable urge transmitted to the conscious mind to trigger immediate corrective action. This is one of the unknown forces of the Universe protecting the viability of the human species on planet Earth, a blessed boon. It is, too often, an unrecognized benefit given to unappreciative sentient creatures in mysterious, temporal time and space. Humans, it is believed, have yet much to learn about the mysteries of the universe.

Readers Space for Thoughts and Ideas.

The First Five Seconds:
Capricious Key to Instant Rapport

The showtime is set. The houselights in the Old Casino Theatre in Toronto have gone down, and the orchestra hits it from the top. In a minute, from behind the right side curtain at the back of the stage, a nondescript man starts to amble from right to left, back of stage to front, with an ego leer on his face as he eyes the house, a matinee afternoon audience. Without slowing a whit, he tumbles ass over teakettle off the front of the stage, deep into the orchestra pit.

His audience is transfixed with deep concern. They watch as he picks himself up, clambers back on the stage, limps sedately up to the microphone, and says, "Who in 'ell tripped me back there?" The audience roars with delight at having been totally snookered, and from that moment on, the stand-up comic (and that was what he was) had them all in the palm of his hand. The first five-second ploy gained him instant rapport, and it served his purposes, to create a warm, receptive audience ASAP. Too drastic for others, but the idea is effective.

Transferring sensation of any kind between speakers and listeners can be a tremulous affair fraught with insecurity and fear of failure. If a successful result is mandatory, the pressures on the provider can be devastating. Hence, the first five seconds takes on a starring role in turning opportunity into success.

In considering the component parts of verbal intercourse, it can be realized that a vast difference exists between idle conversation and true communications. Conversation, once contact has been effected, flows at rates and levels compatible with the comfort of both parties. The intellects involved usually set the limits, finding mutually interesting thoughts to share. The first five seconds need not contain much more than the presentation of a stress-free situation, a realistic face-off, with a point of interest to both parties, that will serve to engage the attention and interest of the recipient in a positive way. This reception will then proceed positively or negatively, depending on the eventual disclosure of the purpose for the contact. The personalities merge into compatibility if the right mood prevails.

In a communicative versus a conversational situation, the first five seconds gains a patina of surreal importance. Concepts of approach and follow-through require some preparation at the subconscious cosmic levels. This is endorsed by the fact that the instigator of the event has a mission to convince the recipient to engage in some debate and to make a decision to buy what is being sold. Reaching a willing ear to listen to the opening dialogue becomes the primary issue that is meant to arouse compliance for the engagement. Regarding the first five seconds, whether it is to sell an idea, a product, an agreement to affect a change, or any other transaction that might take place between two strangers, the initial contact will establish the early mood for the success or failure of the venture.

Recognizing that the recipient may likely react negatively to a new submission, the proposer should then make the initial contact with emotional tools. The much-overused mantra is, "And how are you today?" It is a negative-arousing tactic, creating immediate hostility if the recipient recognizes the vacuity of the comment. Some pre-research on the personal interests of the target subject is vital so the opening gambit expresses a real interest that defines the person and his or her accomplishments rather than his or her status of existence. An example might be, "I hear you are an expert in the field of …" or "I see from your office treasures that you are an expert in …" The emphasis should be placed, right from the get-go, on a meaningful comment that draws the target into a positive net, pleased to be thus recognized. If asked to be an expert, the prospect opens up. It is five-second magic.

Most people appreciate being recognized for their talents. The importance of recognizing this characteristic is as dramatic as the act of tumbling off a stage into an orchestra pit.

Readers Space for Thoughts and Ideas.

Total Commitment: Do or Die in the Valley of Death

Half a league, half a league,
Half a league onward,
All in the valley of Death
Rode the six hundred.
"Forward, the Light Brigade!
Charge for the guns!" he said:
Into the valley of Death
Rode the six hundred.

"Forward, the Light Brigade!"
Was there a man dismayed?
Not tho' the soldiers knew
Someone had blundered,
Theirs was not to make reply,
Theirs was not to reason why,
Theirs was but to do and die:
Into the valley of Death
Rode the six hundred.

Cannon to the right of them,
Cannon to the left of them,
Cannon in front of them
Volleyed and thunder'd;
Storm'd at with shot and shell,
Boldly they rode and well,
Into the jaws of Death,
Into the mouth of Hell,
Rode the six hundred.

Flashed all their sabres bare,
Flashed as they turned in air,
Sab'ring the gunners there,
Charging an army, while
All the world wondered:
Plunging in the battery smoke,
Right through the line they broke;
Cossack and Russian
Reeled from the sabre-stroke
Shattered and sundered.
Then they rode back, but not—
Not the six hundred.

Cannon to the right of them,
Cannon to the left of them,
Cannon in front of them
Volleyed and thundered;
Stormed at with shot and shell,
While horse and hero fell,
They that fought so well,
Came thro' the jaws of Death,
Back from the mouth of Hell,
All that was left of them,
Left of the six hundred.

When can their glory fade?
Oh, the wild charge they made!
All the world wondered.
Honor the charge they made!
Honor the Light Brigade,
Noble Six Hundred!
(Courtesy of Alfred Lord Tennyson, 1870)

The incredibly emotional words of Lord Tennyson speak descriptively of the fears and courage of humankind and shoot hard cannonballs square into the target of total commitment to succeed.

Whether that commitment is personal or professional makes no difference. To grasp and hold on to the brass ring of successful achievement calls for a dedicated commitment to carry through all assignments, whatever their scope and challenge may be. Such a commitment involves a serious look into one's personal depths of courage to find that unlimited, subconscious, cosmic strength and energy to carry difficult decisions to the ultimate goal of success. To put it succinctly, "Once mirrored as a giant in the introspective eyeballs of oneself, a human cannot fail his or her strategic objectives regardless of all impediments he or she may encounter."

Without fail, research shows us that all successful humans possess one common trait when seeking success in whatever form they visualize. They burn the bridges of all easy exits in advance and, firmly grasping their guts like a rope, swing over the Valley of Death to reach the other side. Once launched and swinging free, they entertain no serious doubts but that success is assured. Because such validation calls for powers that may well exceed those employed in normal pursuits, the will to succeed must find extra strength and conviction, perhaps in a connection with the cosmic powers of the universe. Because they are part of that energy, it calls only for the belief that it exists and is available. Most superhuman effort is not needed 24/7, so it is a worthy pursuit to seek intervention only whenever dire events are critical. Time and patience notwithstanding.

Cannon in front of them
Volleyed and thundered;
Stormed at with shot and shell,
While horse and hero fell,
They that fought so well,
Came thro' the jaws of Death,
Back from the mouth of Hell,
All that was left of them,
Left of the six hundred.

To which we would all say, "Amen." Total commitment indeed!

Readers Space for Thoughts and Ideas.

Wisdom: Like Good Wine Over Time

Wisdom is not a virtue with genetic roots. The capacity to learn is a given at birth, but wisdom combines knowledge, experience, and insight. This must be included for wisdom to be valued in the realms of reality.

Insight, which could be the most important of the three virtues, is the ability to use the other two effectively. Because insight is not a genetic trait, it must originate in the subconscious mind and gain its special abilities from the cosmic energy spectrum that surrounds the Earth and all its components. The transfer of this cosmic energy, as required, occurs naturally as the human is already a part of this energy, which is inherited when impregnation occurs.

Knowledge, the product of learning, gains different levels of understanding from all of life's experiences. Wisdom, however, takes a different path to realization. The process of vintaging a quality wine is an example of how the virtue of wisdom develops.

Compare life's experiences with the grapes from which wine is made. As in life, some grapes may be sweet, ripe, and robust, while others are bitter and wizened by unfortunate experiences gained in coming to maturity. To distill the product, only good grapes must be contained in a vat. In life, that vat is the human body and mind. The condition of the container must be sound to hold the juices of the grapes and the energy of the human intact. If the barrel is weak or otherwise damaged, the wine will suffer. If the human has had an unhappy, perhaps violent upbringing, reverses will scar his or her route to wisdom. Pressure on the grapes brings forth the juice. Ambitious effort raises the awareness of humans and will build on their history of living. Properly tended grape juice, mixed with selected ingredients, must then be introduced to the winemaking process and set aside to mature. The human needs the ingredients of teaching and knowledge to start the period of maturing into a lettered member of the human race.

Then the part that turns juice into wine and intelligent humans into containers of wisdom starts. In the process of maturing, the brews start to change. Chemistry invokes its magic into the juice and insight, likewise, into the human. To make both superb, the element of cosmic energy must be inherent in the chemistry, given that all the ingredients have mastered the challenges and time for reflection was attained. Vicariously, like blankets of mist creeping over the meadow, the essence of wisdom and the vintage quality of wine begin to emerge.

One can never truly define the taste of fine wine; nor can one ever portray the essence of wisdom. But once having experienced either or both, one can never forget. Young persons, like young wine, develop individuality, apart from their origins, quite rapidly. In so doing, the fermentation stages can be severe. While sporting images of spirit, color, robustness, and potential, because of the raw edges of immaturity, they rarely possess all the virtues of wisdom.

Access to the elusive cosmic energy that cloaks the Earth, in order to solve specific human problems, is not like turning on a tap. Awareness that this energy reacts voluntarily if needed, but will not react to commandments or requests to emerge, is vital. The key to the door is more psychological than physical. Cosmic energy is not generally accessible to all. It exists for everyone but operates by rules of universal law that govern accessibility. Responsible thinkers can mature from adolescents to productive adults if the mental energy is wisely utilized and the wrecking crew of abuse, excessive early ingesting of harmful substances, and damaging activities are rigorously denied or controlled. Every human owns the right to manage their personal destiny once they pass the barrier of risky adolescent experimentation.

Readers Space for Thoughts and Ideas.

Modern Roaring Woman: Butting Away at the Stone Walls of Male Tradition

One of the more fascinating topics is found in our study of the rapid, even meteoric, emergence of women into the murky worlds of male-dominated creative and business traditions. The divine right that has existed for centuries, fixing male/female roles into hard boxes of steel, has been eroding, and the evolution of changing roles is entrancing to see. How well both sexes embrace the shifting sands is worthy of some broad-scale discussion. On balance, one can conclude that neither sex has handled it very well.

If one deigns to probe beneath the blankets of the obvious, it is quite evident that traditional history has ill prepared neither sex for an exchange of roles. The graveyards of many once idealistic male/female gardens of Eden are littered with the mangled souls in conflict born out of their innate competitiveness. Woman has long prospered using her native, often superior, intellectual wiles, while mankind has clung over long to his illusions of raw power superiority, his physical muscles, and gambling propensity.

In cold analysis, man has never had, in his arsenal, a weapon that could compete effectively with the downright taunting but realistic instincts of woman when it comes to differentiating between fact and fiction. Toss in the towel if she adds the subtle pheromones of femininity.

Equally frustrating can be the instinctive romanticism, the illogical urge that drives mankind to try to grab impossible rings of opportunity as they float, like pale shadows, across the horizons of his imagination. Add to that his visceral instinct to upscale his female relationships from casual to intimate as rapidly as possible, often without any sense of commitment or remorse. It's small wonder that the twain tend to remain like two strangers suspiciously circling their wagons in front of the ever-locked gates of a mutual Valhalla.

In business, one of the real concerns is the tendency for some striving female execs to misdirect their challenges into masculinizing their modus operandi. Similarly, no less disturbing is the tendency by some male executives to hide their personal insecurities behind stone walls and verbal put-downs when they should be supporting the good qualities of their female cohorts.

In the lens of the Macroscope, women and men are seen on the planet to be supportive and complementary, not combative and personality-competitive in business or life. That they are both is illustrated by the volume of separations that occur in those areas where mutual compatibility should work.

It is a sad commentary to have to conclude that the basic, illogical insecurity in both parties prevents many a potential relationship from maturing into richly patterned, productive experiences.

So it would seem that a new sunrise must be evolving. Perhaps that new day will see men and women combining and coordinating their best qualities for mutual satisfaction in all things of real value so necessary in patterns of joyful living experiences.

Women, be they hetero, homo, or bisexual, should perhaps start to ease up on their wars for equality and have a bit more compassion for the males with whom they may live or work. The evolutionary forces of nature have the tendency to ascribe equality well on the way to an effective balance. It requires no more than human common sense to succeed. A force called evolution will help bring balance in good time, if not in current, but during future eras.

There is no reward in the wars between the sexes. If it is acknowledged by both males and females that each has the right to their own life force, compromise cannot be lethal.

Readers Space for Thoughts and Ideas.

Aphoristic Mind Benders: Trite and Tricky

Without some relief from time to time, philosophical works can become exhaustively tiresome and needing of assurance that there is always room for different kinds of challenges. Thus mind-benders have been popping up in the dialogues frequently enough to provide a welcome easing of academic pressure. Test your ability to dissect the concepts.

1. No matter how high the wild goose flies, you can always break a window with an axe.
2. Ticktock, the wall clock, has no face in cosmic space.
3. To stamp angry feet in soft cow dung just isn't the game.
4. The throne is the seat of power for a king and the power of the seat for his subjects.
5. The mind may enjoy all that sitting and thinking, but meanwhile, your arm and leg muscles are shrinking.
6. Gazing at statues in marble is fun. Have you ever considered what it's like to be one?

7. Of many frustrations for man on the make, the biggest of all is the female headache.

8. Tactics win battles, strategies win wars, concepts shape humankind, and philosophies shape the world.

9. While digging in the earthy dust of destiny, humans find their footsteps on the moon.

10. One philosophic idea makes the mountains rumble. One man's muscles might perchance upturn a stone.

11. If linguists suggest that trite phrases should go, the first two should be "They say" and "You know."

12. Don't tread on the tail of a rattlesnake just to shut off the noise.

13. Moon-seeking kites will run out of string.

14. Yesterday is history, today is opportunity, and tomorrow is a prophesy.

Readers Space for Thoughts and Ideas.

Fate and Destiny:
Two Forces Handling Human History

Of all the dreams that human beings wish they could fashion into experiences of reality, to gain a lucid picture of their life as it unwinds into the mists of their tomorrows, the one that takes priority over others is the road of life called destiny.

Destiny, not a structure or framework that can be viewed in solid form, is a concept that attempts to crystallize the idea that each person has a singularly defined pathway leading to a full and satisfactory life. Because it is so unique to each individual, the road of destiny defies the easy answer that "one size fits all." However, writings that attempt to make that claim fill the pages of history. These wanderings into abstract psychobabble, while sometimes interesting, in the main only serve to confuse the issue of what destiny might be and how it differs from its own blood brother, namely fate. By any rule of conduct, it must be acknowledged that trying to directly teach another soul how to live according to defined precepts of philosophy is whistling past the graveyards of frustration. A free spirit may be constrained but never killed outright. Can one then make some sense in trying to ideate a viewpoint at the levels of comprehension well below those with the mentality of one Hercule Poirot (Agatha Christie's shamus savant)? One dares; one dives. Not to lecture but to challenge traditional thought and encourage the fires of creativity in someone to flame up with new and innovative concepts.

With intrepid imagination, it is possible to see destiny as a roadway that commences to emerge from the moment that sperm impales the defenses of the ovum and terminates when the soul energy departs the body corporeal and vanishes into the universe. The process of creation that takes place is a marvel in itself. The union of the two elements creates a blueprint that details a new human and draws energy from the massive stores that exist within the caverns of the cosmos, so it gives impetus to the new creation to develop according to nature's immutable blueprint. It takes two elements to create new life. The ovary and the testes, and the systems of reproduction matched by all living organisms, must unite the output of these organs in order for the forces of creation to proceed. The basic prescriptions are normally the same for all, but extenuating factors are involved to make sure each new entry has its unique characteristics.

In that blueprint are all the designs of body parts, their size and composition, and the time and space they will occupy as each entity is created. However, in addition to those observable components, the blueprint also carries with it all the genetic markers and tools of survival, along with the motivational, intelligent, and emotional factors and their inclinations. These birth time facets of a human contain, as well, the genetic markers that incorporate the basic characteristics of all ancestors and changes to the natural pattern. From these suppositions, one can presume that the basic formulation of a personal destiny is what the road ahead should be for the new entity, but only if the forces that are playing the creation game can do so without pressures of change from the human's personal environments.

But, of course, that cannot be. From the moment that the newborn infant emerges into the earthly world, external attention, coddling, and control encapsulates it. The ills and chills of the surrounding ecosystem and its Homo sapiens start to influence its physical and psychological environment and, in so doing, will constantly challenge the natural size and scope of its destiny.

This brings the factor of fate into the picture. With a rather simplistic definition, fate can be said to be the events that finally define what one's destiny will actually be and not what was called for in the original blueprint. While elements of fate cannot divert the basic outcome, they can introduce new forms of energy, both good and bad, that may or may not affect a change in the scripted focus of one's destiny. The two known and certain points here, which are immutable, are progenitive insemination and quietus. If one is born, one must eventually die, naturally or otherwise.

However, the linear time period of one's existence is variable and can be extended to its natural limits or truncated in so many ways during the experiences of living.

If one starts with a manifest destiny, then it may change because of extenuating circumstances. Accidents of all kinds can occur without rhyme or reason and can change the intended path of destiny. Diseases with short or long-term impact may cut the blueprint forecast of a life expectancy without warning. Unlawful, self-imposed excursions into societally aberrant behavior may cause serious devolution in the use of one's natural talents. And a poor and devoid-of-affection home environment can attack the psychological ambience of a human to such an extent that their blueprint of destiny is seriously altered or utterly destroyed.

While the roadway of destiny is somewhat fixed at the time of birth, then it can be suggested that fate can alter its variable facets. It has been proposed that fate and destiny are one and the same. However, while they may share the same time and space, they will always be separated in the human mind by the influences of the unplanned and spatial energies that constitute and affect the body corporeal of those creatures we call Homo sapiens.

Readers Space for Thoughts and Ideas.

The Big Me:
Frozen Philosophy in a Frenetic World

Perhaps, in one form or another, the concept of a self-serving interest has been a part of the human psyche for untold time. Pure survival, intensive competition, and a subconscious with cosmic overtones seeking individual recognition in the human mix-master of living have fertilized the soil of a need for personal independence sometime long ago.

Looking into the reasons why the increase in the inward-looking characteristics in high profile individuals is evolving so dramatically, some facts become obvious. Whatever the human instincts that are dominant in their decision processes, two basic attitudes prevail.

Human character being as ubiquitous as it is, the one attitude that seems to dominate is the desire in one human to exploit or use other humans (employees, friends, and associates) for his or her own aggrandizement. Such personalities have little respect for their cohorts and use the absconded talents and productivity freely with no thought of passing on credit to those being used. This style may be considered as bossy, and its possessor takes criticism unkindly and may even take revenge on its critics.

While this type of manager exists in many concerns, his or her employees may detest him/her, and work may be subtly sabotaged as a retaliatory measure. In some cases, such employers have a low understanding of the fact that people are not machines and they do have feelings. If underlings pose attitudes that imply or state that they are selling only their services but wish to retain their own souls, perturbation may be inevitable. The natural force of survival may engage and add another dimension to the operating strategies being employed.

In studying the nature of "Big Me" types, deep insecurity can be found in their self-evaluations. When they pass through the doors of fame and fortune, it conveys a vision of a golden sunrise and a cloudless, blue, blue, sky. If this image does not translate from fantasy to reality, the notion of importance may translate into a somewhat regal reaction, and the role assumes the trappings of supremacy. That, in turn, infests a mind that lacks a real sense of logic with seeds of values that, like weeds in a flower bed, will grow to be those plants out of place. Insecurity will undermine the rapture of success in time, and the path through the weeds must suffer the slashing blows of the machete to ensure the status quo. Of course, the slashing strokes do not discriminate between the genuine plants, who do the real work, and the invaders. Real control of the situation moves from positive to negative, and the organs of productivity slide down to jeopardy.

There is no sweet spot in the language of equality among humans if the shape of control fails to emphasize the virtues of logical respect and compassion by all for all the sentient creatures on the planet Earth.

The second side of human development lies in the ability of those with a mandate from nature to govern, seeking new perspectives in the management of the incredible energy that is constantly foaming around us. This calls for a change from the "Big Me" philosophy to one of the "Bigger We." Not an easy option for a type A personality. However, with some cosmic re-evaluation, boss types can effect a change in their attitude toward those who depend on them for planning and control.

Cavemen, when survival was their most prominent challenge, did learn over time to understand the virtues in group unity. A family concept evolved, and from that grew the fact of community life. Community in more modern times evolved to villages and, from that, matured rural/urban dependency with countries, nations, governing bodies, and world co-operation as the basis for the concepts of an idealized multicultural amity.

The "Big Me" concept, however, due to the greed for wealth and power that is endemic in the souls of humankind, has exerted continued influence on world ambitions for equality in resources and lifestyle itself. Wars, which Big Me autocrats think to be the equalizer, fail to fulfill the promise of a global resolution through melding human problems here on Earth. Perhaps it should be recognized that the laws of nature, set when the universe matured, never did intend that the polyglot of races should attempt a melding. The idea that the world we live in should become one unified and happy place is, after all, an idea that has come from the minds of humankind ill prepared to visualize the disastrous consequences from such ambitious aims of unity under one control.

At a more comprehensible level is a fact that every force on Earth contains a counter force. For that reason, the Big Me concept fails to gain any dominance in general life and its tools of commerce. The power of the boss to enforce change that does not consider the possible turmoil, if the changes sorely rankle those affected, will fail. Enough failures should beget more comprehensive methods to teach, support, and encourage cooperation in the interests of productivity through mutual concentrations of energy, materials, and proficiency.

Too often, the visions of power-moguls transfer focus on their own big picture. And if that prospect narrows, so does the possibility of success. The bones of lost opportunities litter the graveyards of many titans of industry who failed to abandon their Big Me ambitions for a more humane sense of productive motivation.

Sharing the richness of resources is denied to human greed that has no limits.

Readers Space for Thoughts and Ideas.

Ego-Theism: Adoration of These Feet of Clay

Once upon a time, a Solomon reincarnate, in mists of wisdom, said with some authority, "Power corrupts, and absolute power corrupts absolutely." This prophet must have had some cosmic insight into the very souls of humankind. His words are polished and proclaimed in modern crucibles that occupy our daily economic and social lives. For most humans, each has one destiny that may take them down many detours and leaves them with emotional overloads that require subconscious strength to enable them to meet the challenges.

Without that infusion of extra energy, how could humans possibly control the more savage instincts that seem to surface when the rails of absolute power on a lordly scale lay out on the roadbed of their fate? How is it possible to keep the locomotives of lust, greed, jealousy, spite, and hatred from accelerating to disastrous stages on the rails of life? How does one cure deafness in the presence of valid opinions expressed but disrespected due to the blinders of vanity?

The basic reasons for the corrosive impact of absolute or perceived grandiose power are many. There is reason to believe, qualified to some degree by long experience, that a basic reason may lie in the inexhaustible desire by humans to be deified by their peers. Not that they would admit that to themselves, but it is a characteristic that seems to be inborn to be the superior among competitors, a water-walker in a soaking sea of humanity, and the big winner in the fields of also-rans. Perhaps the idea of embracing super importance to one's ego is a natural development as the competition for success rides heavily in the milieu of increasing competition for the rewards from the idealized importance of fame and fortune. Opportunity for some but only dreams for many.

The fodder for the nourishment of ego-theism is found in the sense of hero worship that enthralls the masses. Apparent success by selected idols is seen as worthy of another's idolatry. Perhaps it gives an aura of power that might transfer to the adulators in some magic way. A cosmic transfer perhaps?

In logical examination, however, such adulation only serves to wrap the egos lightly in the fragile tissues of a vulnerable self-respect.

Human power engines cannot be said, in truth, to be designed and built entirely by the vicissitudes of experiences. Because vanity is found to be a force of such great magnitude in the spirit of humankind that it will subjugate the counter force of pride, small wonder that so many ego-theists wind up, in life, adoring their own feet of clay. Ego, serving as a motivator, rarely considers the reputations of those it shreds, nor those it may sometimes glorify. A worthy subject to give to it some deeper thought.

Readers Space for Thoughts and Ideas.

Tortoise Travails:
Trampled to Death in Hare's Race to the Top

In the era of historical truth, the tale of the hare and the tortoise suggested that the slowest appearing one would be, in the end, the first past the post. Labeling the tortoise as experience and the hare as academic knowledge, the current picture today would favor the hare.

There was a time when being unqualified for a position of prestigious merit would have been a lack of a Harvard (or comparable) degree. In current terms, MBA qualifications are a pass through the gate of business success, at least in commercial worlds, and PhDs are no longer a rarity in other professions where merit is revered.

With politics as a bellwether example, those with higher academic credentials or professional reputations have mainly held leadership in this field. Even those tools, however, while they seem to produce members with consummate skills in grasping the big rings of power, do not prevent a drastic ineptitude in handling their tasks once in office.

Could it be that the lack of ground-level experience leaves many pro politicos vulnerable to bad advice and snow jobs with such perversions of truth slipping through the transparent veils of subterfuge? Tribes of outside legal eagles and consultants only reinforce the belief among critics that the hosts of such services do not know what is right, nearly right, or dead wrong with those decisions he or she may be called on to enforce.

In the business arenas, where the devil Failure sits ever on the edge of the executive's desk, the costly gamblers in the academic wonderlands have begun to realize that the big myth of academic superiority in all things is not what it seems on the surface. The illusion has been that such degrees guaranteed a neat, hassle-free glide to the top of the temples of power. After many disasters, the major executives are beginning to see that experience is weighing into the mix and is tested best at the foot of the hill. Here, the toes of all levels of competence can be found, and the carbuncles are tender and must be cured on that escalating slope of ambition.

So like the phoenix, the tough bird of experience is back out of the ashes of disrespect by mentors and practitioners, and some basic truths are apparent, dead center in the minds of caring, thinking strivers for power.

Theories are necessary in the stages of planning and ideating, but the hard flame of experience must be at the head table too. That ensures that all those academic plans are logically executed if the expenditure of time and resources are optimized to assure a true measure of progress. Experience, however, needs to be communicated to others with the tools of empathy and true modesty.

When it comes to rating the equality of the race between the hare and the tortoise, rules must be established. There would be little objection to the concept that both academia and experience, when they are in a mutual harness, are the best team to move events forward with some modicum of an assured success.

The captain who has his hands on the tiller cannot too often be aware that his ship may not fit into the canal smoothly if he lacks the experience that governs the fine points of that challenge. Perhaps some understanding of this omission in expertise is why special professional pilots are taken on board to fine-tune the passage through the critical phases. This rule is sometimes ignored, and the risks of disaster escalate.

The rungs on the ladders of success are generally well spaced in truly professional organizations. One does not fly test planes without his or her logbook filled with hours of airtime performed and noted. The pilots, in hours of time spent riding the ozone in test vehicles are paid for the experience expected to be expounded in their curriculum vitae.

To become a talented carpenter or plumber mandates field experience in evaluating the problems and experiencing the tools of these trades. Trade schools can introduce the basic theories that are entailed in these professions, but apprenticeships in the working milieus are mandatory to gain the essential credentials. University education is a lesser virtue on the totem poles of the serious practitioners of tradecrafts; remuneration is highly paid for the experts with tortoise qualifications.

Readers Space for Thoughts and Ideas.

Galactic Revelations:
Data Bytes from the Great Beyond

Whenever any person has the audacity to challenge the commonplace by thinking beyond the ordinary rules of general discourse, it is often said that he or she has lost his or her mind. The idea that he or she may have found his or her mind is thereby called pure nonsense.

For many reasons, the negative POV (point of view) seems more typical than a positive outlook. Behind it all, perhaps lays the basic human fear of being wrong or having to deal with an unknown. Pragmatism, a major human attribute, may naturally cause thinking that opinions formed outside established rules are asinine, unproven, and therefore unacceptable.

Coupled with this attitude, the human trait has a basis set in the belief that those whose ideologies are beyond the average norm deserve to be chopped back to meet the standards. Philosophic thinking is considered the law in academic spheres, but has less prominence in ordinary thinking, a conundrum in human attitudes.

But think. Without mental challenges that slop over the edges of traditional platforms, there would be no real progress in human development. The laws of evolution deny such rigidity.

It takes courage to defy the mass mentality that prefers the comfort of established norms. This defiance is thought to affect the normal stability of radical reformers and casts them into isolationism from the herd. But highly focused specialists are seldom social butterflies. They are introspective by choice when profound mental gymnastics become an obsessive preoccupation. This sets them apart from the society where commonplace is the comfort zone of humankind. It can be accepted that communicating in the cosmic environment is beyond the dialogue of many occupants of planet Earth.

Examples of the progressive rules of change in the lives of humans on Earth are many and consistently changing, if not always improving the environments of existence. Most everything that humans enjoy today started out as an idea in a single, curious mind some time ago. Many more new ideas are being hatched to-day as time marches humans into the evolutionary future.

Since proof of events occurring in the void of space and with the Universe as an entity, much of the information that is published has a generous larding of speculation involved. That should not deny the presence of thinkers in the field of speculation when it comes to thoughts about the events in outer space. The ideas of a logical lay person can have as much credibility in examining spatial phenomena as the experts if the ideas are tossed into the roiling mass of both expert and free-based ideologies for tests of their grasp of cosmic realities.

Readers Space for Thoughts and Ideas.

Nationhood: Tic-Tac-Toeing on the Battered Belly of a Pregnant World

What is the concept of nationhood? In the minds of politicians, concerned ... nay ... obsessed as they appear to be with causes and ignoring, by design, any inevitable effects and consequences, it means easy, cushy, patented formulae for everything.

Economists, toying with their computer models, see ideal answers in the mathematics of their recommendations but are unable to program the idiosyncratic influences of humans into the mix. So, in all cases, the human factors may be ignored or denigrated.

In the minds of average citizens, there are simpler realities. They are:

- Having a piece of planet Earth to call one's own but decrying the eternal irritation of having to pay for its use over and over through abusive taxation and calls for financial supports.
- Not being burdened incessantly with threats to one's personal security

- Not having to face ugly prospects such as the lack of earning opportunities.
- Not having time for relaxation while seeing others with these benefits
- Not having rules that intend to micromanage private life and prospects
- Not having exorbitant use of the same privileges as mandarins in power who may have lost somewhere in history, the real reason for their existence to be responsibile to the public.

Nationhood cannot be stuffed into a box of theories. It is made up of individual humans with personal agendas and often with obsessive loyalties to the place where they were born and where they grew into maturity. The concepts have to be, therefore, pragmatic, idealistic, philosophic, and elastic and contain all the hopes and dreams of multiple genetic human entities.

For these reasons, the controls that must govern how a nation is defined to its peoples and the surrounding world cannot be the product of some idle meditation by incompetent purveyors of their own idiomatic ideas. The magnitude of both scope and limits demands the thinking of humankind beyond the principles of personal ideas based on unreal platforms of reaction. Deep introspection by competent minds might even experience difficulty in imagining the aftermath of convoluted mental proposals. The employment of subconscious cosmic allusion would help in the process but is often not part of the process when time pressures are assumed. Inveterate ideas and logic, submitted for consideration by uninvolved minds, tend to divert the focus of constructive ideation when the task is to establish clear meanings in defining a philosophical platform for nationhood. Too many cooks tend to water the soup, especially if the cooks are thirty-minute managers.

This loss of control to gargantuan centers of power, assets, academic credentials, and general influence can send the central concepts of nationhood spirally off into wastelands of unrealistic disparity. The essence of meaningful development, when the task is to establish a concept of nationhood that satisfies a maximum number of subjects, is lost. It begets the negative forces of non-communications, misunderstanding, and dispersion of critical responsibility. The natural results are feelings of helplessness, apathy, and withdrawal from active participation in the building of nationhood.

Earthly humankind, as mirrored by the majority of ordinary citizens, could generally understand the scope of working a family unit when it comes to the rules and controls that are needed. Beyond that, they may project upward to the next level of community. Above that, they must try to understand regional and then national involvement. From that point on, they may fail to fully understand the dynamics of continental units or, finally, the planet and the universe.

When comprehension dissolves to the more localized sites, the rest may be seen as remote specks on the brain, more frustration than facets in the general concepts of birthing a comprehensive state of nationhood.

Nationhood, as it is generally apprised, is viewed as somewhat like a giant structure that is cobbled together with a mix of components assembled in some uncanny way to represent a homogeneous whole.

If given enough time and left to evolve without excessive tinkering by inexperienced or uncaring mechanics, all the parts do somehow tend to merge into a harlequin fabric of nationhood imprinting the landscape of history. The tendency of setting loosely connected elements into a concept or ideas of perfect totality without logic soon enters a phase where the unseen power of balance will move the disparate parts into a recognizable whole. But that may take time and patience, and these force pressure decisions that are generally accommodations to the need for immediate solutions. Patience has no virtue here.

To understand the concepts of establishing a balance in forces, one must consider the influence of evolution on all factors that affect structures, events, and attitudes in general. Evolution, being an immutable and cosmic force, has the power and intent to always retain the need to restore normality. Normality, under cosmic law, means a return to the balance of all elements vital to the orderly progress of events in the universe. Planet Earth is essentially bound by these rules. Hence, nationhood must be defined in cosmic terms to survive.

Nationhood requires only a major container of patience, a barrier to keep irrational anglers away from ponds of vocal permissiveness, and a society that believes in the concepts of mutual respect and community fellowship.

Readers Space for Thoughts and Ideas.

Honesty: Ragged Doll on the Cosmic Stages of Power

Honesty is a virtue that is under the personal control of each human on the face of planet Earth. It has no universal definition, but because it is an acceptable reality, its value is in the intrinsic cloak it wears with most people.

Truth is a virtue with roots in nature and has immutable value. It lies at the base of honesty and is the way that those who value truth see it as the yardstick on which honesty is measured. In fact, it is the ruler on which the essence of truth becomes a variable in judging the probity of humans and is therefore ultimately married to the cause of honesty.

If one argues that honesty is a permissive, nebulous attitude, then truth takes on the raiments of steel-coated facts. Facts are based on provable actions or events and represent reality at each point of consideration.

Honesty has no cosmic platform of granite and is therefore subject to its use as a tool by liars and truth-tellers with equal rectitude. It is used as a costume to conceal the real meaning if the liar chooses to believe that a lie will suffice to ease an event or situation through the barbed wires of public and personal observation. The motivating force, of course, can be found in the human tendency to avoid responsibility if the truth appears to be penalizing or otherwise uncertain. It has a countervailing strength in that honesty might be used as a guarantee of validity if the factor of proof is under consideration.

In the business communities, a third virtue can be and is used to arbitrate the space that exists between honesty and truth. This may be identified as ethics. Use of this value calls for examination of the other two usages for their coloration of feelings and attitudes, that is, when and how any answers must be found. The spontaneous lies may be uttered without recourse to proof of their validity and can become a shadow of both truth and honesty. Introspection, the personal monitor of both honesty and truth, is the thinking that is almost always present in critical stages of any discussion. The need to look inside one's mind to limn motivation can be redundant if the thinker is a congenital liar or inveterately opportunistic in gaining his or her ends.

In the political arena, honesty can be and is a fleeting wraith if ideology, with far-reaching consequences, is to be protected. Dealing with a wide spectrum of humanity is difficult in seeking a positive consensus, and the easy route to some success, in that endeavor, is intelligent obfuscation. The fact that sharp-eyed publishers in the public media often bring such methods to light is no barrier to its frequent usage.

Discovery is often rebutted with bluff and bluster. Challenges are a provocation of ill feelings and emotional hatred among those with a heavy onus of belief in personal power and position.

Governments are not destined nor structured to be dishonest. They are built to be purposeful, useful entities whose mandate is to create environments that enhance and enliven the rights and privileges of the populace that has elected them for that specific purpose.

However, politicians can be dishonest, evasive of the truth, and self-serving to the Nth degree. In an environment that is wide open to corruption, those members of the system are quite free to exploit their host and let their loss of integrity lead them down greedy avenues. If the leader of a government has strong ethical beliefs and tries to clean out corruption, his compatriots and their partners may revile and stultify him with perfidious tools. The leader is soon seen as a traitor to the group. Using their wide-eyed, innocent tools of media and paid hacks, they will define him as imperfect in some way or autocratic in another. No leaf lies flat; no stone is unturned to find minor deficiencies of character that can be blown up or falsified. The opposition is electrified, and using the gullibility of the grandstand with its visions of corruption, they will press the buttons of their bogus integrity to slay the ones who dare to challenge their viewpoints, security or presumed invulnerability.

If one thinks seriously on it, who calls politicians to the witness box? Not the systems that exist within the organism. Not the public who are led to believe that elections serve to clean the house, but only change the cast. The real power in a democratic system of government is with the non-elected bodies who maintain the operations, the so-called civil servants. They remain in power while the elected politicos may vanish in the polls. In the United States, key public servants depart with their deposed bosses. No doubt a system to mitigate entrenched opportunism. Whether it actually does rein in corruption is rarely examined.

It is said that, if a newly elected head of a government department does not conform and prolong the privileges of the non-elected public servants, these same bodies have the position and power to smother the efforts of their party leader. As far as the public, to whom the government is beholden, is concerned, they see a few barn doors closing after the horses are stolen. Gone and soon forgotten until the next theft occurs.

If there is any justice in the world of honesty and truth, it exists in the fact that evolution tends to even out the bumps in human greed and drives for power. In this harried world of wolves and foxes, the visionaries of true justice may relate to those dancing Gay Gordons on the public greens and point out that no one human can live for long enough to seriously abort an immutable truth. Perhaps the intention of the universe has mandated only short-lived destinies for all humans in order to mitigate the damage that might be visited on the hearts and souls of the planet Earth peoples over longer periods of time.

With most decent, hardworking humans, the rule that sits deepest in their subconscious souls is the time-tested axiom, "Honesty is the best policy." If it keeps the conscience clean, comfort in living productively and long is assured. One wonders if this has not become a hoary cliche that has lost its relevance with modern humankind.

Readers Personal Space for Thoughts and Ideas

Parental Paradox: Sowing Cosmic Seeds of Wisdom upon Fertile Soil

The parent-offspring relationships in the last quarter century have suffered incredible eruptions, shifts of power, and the trials of communicating. Several causes are seen to have been the major motivators of the changes:

A. Evolution in the planetary landscape has created a sea change in the intelligence of the oncoming generations. Children are maturing much earlier and learning to take or wish to take umbrage in their personal views of family rules and regulations. Parents often mired in the mud of old traditions have suffered the loss of confidence when the derisive attitudes of their offspring blow their decisions, once etched in stone, sky high. Youth also places minor weight on consequences, if they are considered at all.

B. The all-pervasive influence of modern electronic technology has created new rules of behavior and more permissive regulations. Professionally tailored by designers of these rules, specifically aimed at younger and younger age groups, these have brainwashed the uninitiated and naïve youths into believing that the lives depicted in their portrayals are the real thing instead of total fantasy with misleading consequences. This catering to young imaginations sets the stage for hero worship that totally destroys the lessons in credibility that a parent may wish to convey. The clash of wills creates the scenarios of family wars and dysfunction.

C. The extensive growth of single-parent households over recent years has broken down the family concepts of teaching logicality. Once family ties are broken, the family compact may never recover. The incidence of young male delinquency is known to be the product of dysfunctional and erratic family life where the father figure has never been involved or present. Pregnant, young single females, inexperienced and disaffected by parents and society, are unable to cope with parenthood problems starting in the preteen years of youth. If such mothers are minimally educated and financially crippled, the single parenting route is deficient in too many ways to ensure a stable, positive environment.

To defuse the intensive impact that commercial and social promoters have on the attitudes and reactions of young, credulous minds, the parental challenges must be both sensitive and positive. The agenda of good parenting begins well before the birthing process is complete.

A. **Family Atmosphere:** From the first moment of birth, nature has programmed the newly created entity to absorb cosmic energy and environmental information. Very hungry minds are drawing in the atmosphere of welcome, the energy of affection, and awareness of the vibrations of all extraneous events. While the atmosphere at birth is vital, it gains importance as the growing desire for information escalates. Family atmosphere, which the parents establish, is created by their feelings for each other and the life that they have crafted. If mutually strong and well vetted to clear out the negatives that may have been visited on each partner by their parents, they then can face the challenges with clean personal slates and all mutually agreed-to compromises that must grace their married union. Parents who are comfortable with each other can, without excessive emotional trauma, create the proper atmosphere.

B. **Parenting Attitudes:** In this definition, the art of compromise becomes a critical component. Human insecurity, hidden and defended, if maintained and expressed in argument and anger between the parents, will defile the air and be lodged instantaneously in the vulnerable mind of their offsprings. Disagreements about tactics and strategies of parenting are soaked up as if by blotting paper and become a part of the learning curve for siblings from day one. Such divisive dialogue becomes their wedge that can be used to gain favors from one consenting partner against the wishes of the other anywhere on the road to maturity. It should be shown in parenting manuals that adolescents do not willingly pay much, if any, attention to mandated orders such as, "Do it because I say so" or "Don't do as I do … Do as I say because I am your parent." The adolescent mind learns mainly from example and less from dictatorial mandates. Most of the values that will stick in their subconscious minds will be those learned from experiences they will have gained from watching their parents mostly, but also from relatives and friends. Exemplary behavior by parental models is most critical in teaching the etiquette of life. Smoking, drinking, swearing, and other negative youth-learned habits cannot be objected to if the parent or parents set the examples by their mutual deportment. However, good habits likewise should be constantly demonstrated. Compassion, integrity, open affection, respect, and cooperation are a few examples.

C. **Defining the Real Values of Life:** If parents have no comprehension what real-life values entail, they may need help from experienced tutors themselves. In the highly diverse environment that has developed in democratic countries, real values have been subsumed for reasons of personal or commercial gain. Those with profit maximization as their primary objective may have no compassion with the consequences of their promotions on young human psyches that may have been distorted and misguided. Therefore, it is mandatory and well credentialed for parents to counteract these influences with tactics of their parenthood. Set good parental examples.

1. The music industry is one example where the ravaging of adolescent minds is most rampant and hero worship is a created phenomenon. Artists who are mostly untalented but naïvely permissive, blinded by the promises of fame and fortune, are used until their promotional appeal begins to wane. Support by their exploiters is dropped posthaste, and newly fashioned icons follow on the stage, presented in spectacular fashion. The name of the game is turnover. Talent, real or imagined, rolls in and out of the music-creating machines posthaste, directed by exploitive demand. It is a well-recognized strategy by product marketers to keep customers in heat by promoting the newest, latest thing. Discs and

downloads proliferate. Wise parents should display their disdain for such lures and explain the negative side of these tactics well before their offspring takes the lures into their souls as some defensive modus operandi.

2. Family commitment is perhaps the strongest tool that parents are privileged to use in communicating the good and bad vibrations to the world of adolescence. Organized events, where both parents and siblings can participate and compete, can take place around a regular Sunday dinner table, for example, riddles, poems, interesting events, show and tells, or other specific projects. Such activities encourage the understanding of respect, parents for their offspring and vice versa. It also confirms the affection that cloaks both and is a beautiful garment worn well in happy homes. The setting will establish the home environment as the best place for siblings to find peace, self-worth, and safety. The birthplace home where the maturation process gains most of its sound values should become a lifetime security outpost for the inheritors of their family dynasty. It can be the best school in behavior that any school can portray.

3. Parenting is a natural miracle. It can never be taken as a sometime thing. Human life is precious. Its

creation should be sacred and its responsibilities well presented.

4. No sane human would suggest that the urges to procreate should be legally stultified or governmentally controlled. However, perhaps stricter penalties might help to reduce the incidence of birthing that is void of total responsibility. If current trends continue to escalate, the consequences that affect the fabric of the world society will overwhelm any logical solutions for reform down the road.

Readers Personal Space for Thoughts and Ideas

Macroscopic Topics (No. 16)

Reflection Time: Tangling with Some Low Points

The Moody Blues

<+><+><+><+><+>

Out there in space, the nightingale
Enjoys the sun-washed hues,
As evening shadows tint the Earth
With reds and greens to choose.
The singer of all nightfall themes
Is captured by the views,
And from her golden throat emits
The notes of Moody Blues.
When captured to expand this choice
The wily bird explains,
"My spirit is caught up tonight
With what Earth's life attains."
Her soul, she said, was much too bent
By feelings rich and rare,
Yet troubled by those down below,
Who do not seem to care.

"I warble in the nighttime air
With notes so clear and true,
But when they echo back to me,
It's always Moody Blue.
Those who hear my music rise,
Perhaps are troubled, too.
But surely they feel more of life
Than sad notes, Moody Blue."

Okay, we'll cheer up.

Readers Personal Space for Thoughts and Ideas

Cynicism: Huge Pebbles in
the Smooth-Flowing Rivers of Time

As the years pass by and one feels life forces draining away into the bottom half of the human hourglass, the pressure to examine the validity of some of life's values increases geometrically. Changes in perspective, attitudes, philosophy, and motivation are viewed in practical, if not jaundiced eyeballs. All the while the hail keeps pelting down to disturb one's equanimity.

Cynicism is an attitude in those who believe people only do all things from a selfish perspective. Skepticism, in the same ballpark, is doubt or disbelief regarding the opinions of others. Both are attitudes that mature over the early years of lifetimes and are born of personal experiences with humans and events in the world. Because attitudes are the product of basic feelings that may have genetic shadings, the influence can be major or minor as it gains some traction over time. Bitter adult cynics usually grow from a dysfunctional youth with a history of physical and/or mental abuse.

Cynicism is a free agent in the human conscious mind and may be a constant that could affect its possessor in all acts of communication. It puts restrictions on the type of careers that demand objectivity because its proper place should be to evaluate events as they are viewed or heard. If the judgments that may ensue call for blatant criticism, cynicism could overstep its limits in each case. This aberration in the process of human dialoguing can create untold friction and ill will. Cases in point are film and artistic critics who permit their cynicism to outweigh their judgment. They find no warmth in their relationships with artists, musicians, sculptors, writers, and politicians. At times, the bitterness is volatile and destructive.

Skepticism, however, is the critic's super weapon. With its milder impact on its recipients, it can be a free-for-all principle in a hearty discussion for the different views that may prevail. It does not have the bitter finality that cynicism can bring to any exchange of attitudes.

In a world that is filled with humans who may take illogical pride in their personal creativity, or who raise brick walls against criticism of anything they do or say, conversing can become a trying ordeal. A defense against the loss of objectivity by viewers or participants can become the seedbed of cynicism. Then milieus for dissent and dissatisfaction are manifest.

True cynics are not born but made. However, of interest would be an analysis of how the mind of single persons can be converted by environmental factors and what powerful forces are applied to make the conversion occur. In general, cynics arrive at their extreme points of bitterness at some evolutionary stage on their pathways to maturity.

Extreme frustration, carried like a backpack over longer periods of time, can bend the will into a negative frame. Such pressure on the will of humans cannot be denied. It bedevils the soul like any other addiction, with the same intensity and loss of control.

Total loss of self-respect, perhaps etched into one's character from a very early age and subconsciously, or overtly dominant from then on, will gradually build a lifetime attitude of self-hatred. That can kill compassion and weave a cloak of cynical judgment that can really never be discarded.

Loss of confidence in and respect for those who wield major packages of power and influence over environmental conditions will gradually convert a skeptic into a cynic, and that may have virulent consequences. In political venues, regular elections can serve to mitigate the emotional responses to an individual's political status quo and hence that of his or her attitudes. But once created and sustained over longer periods of time, endemic cynicism will certainly prevail.

Once deeply ingrained, there is no cure for cynicism. It is a ruling that settles into the subconscious mind, and once there, will command one's conscious awareness to react with negativity throughout the balance of a lifetime. The cure for cynical attitudes cannot be promoted as a universal rule. Professional practitioners such as psychiatrists might effect a cure, but only if the patients undertake a deeply introspective viewpoint of themselves. In truth, a die-hard cynic has little chance of recovering a neutral balance before it is too late.

To be a die-hard cynic means to miss so many of the finer things in life. The tendency to resent and detest people and events that make up the social fabric of civilization would mean that the opportunity to meet and socialize with interesting personalities would vanish. The cynics very existence would then be crammed into a small and bitter package of unhappiness and loneliness. Not the best way to enjoy the few very precious years of a unique lifetime.

Readers Personal Space for Thoughts and Ideas

Progressive Planning: Tuning Out the Fiddlers Stroking Down Trend News

Most standard planning is based on extending the historical experiences plus the current situation into a six month or longer blueprint of anticipation. This presumes that the environment that contains the event on which the forecast has been based will not change in the meantime.

Because our concept of time waits for no man and evolution is a cosmic monitor, recognition of their combined impact on plans that have been validated can be very emphatic but often disastrously negative. Status quo in planning is a giant killer in an evolutionary program.

Despite these rock-hard facts, planners still cling to antiquated tools in establishing guideline recommendations to justify and give reasons or excuses for maintaining regular, historical strategies to drive the planning of future productivity. Records of past results, established formulae to arrive at future numbers, and research as to behavior patterns are all used with some level of confidence and incredible hope for success. For steady periods of stable growth, these tools are passable but lacking in real perspective suited to long-term future needs.

Governments are most often guilty of their predilection for historically based forecasting. Politicians, elected to represent their constituents, often arrive at their posts with no experience or awareness of the scope of their mandates. They rely upon the lifelines from the non-elected civil service to try to gain knowledge about the status of their sectors. Depending on the experience and professional credentials of these advisors, the reputations of the politicos could be sustained. However, once elected and in office, politicians soon discover they are mainly figureheads, whose role is mainly as the public voice for the unelected civil services. Creative thinking from the minds of politicians elected on a platform of personal popularity and no real experience in politics or world-scale perception, is rare or biased. They may have neither time nor inclination to bend their minds to the requirement of serious thinking. Hence, the governing force is not the one the public is led to believe is capable of planning the public agenda.

In countries where the constitution is a republic and where the public service personnel are discharged along with the politicians they represent, the planning functions must perforce be largely based on historical information. In such cases where a president, with his staff, proposes new and hopefully creative legislation, it cannot become law until the congress (representatives and senators) approves the recommendations. Here again, the political leaders are hamstrung by the system that may be archaic and disdainful of the changes taking place in their offices, and on the status of their mandates in office.

In systems where the role of planning is designed as an instrument primarily to control the capabilities of organization, it fails to recognize some basic rules. It overemphasizes the importance of financial objectives and ignores the principal purpose of any important organization, which is to guarantee its long term viability, validity, and its mission and reason for existing. Hence, such planning is directed to the near term, at its peril.

With evolution, a cosmic force that governs the conditions of humans, organizations, governments, and communities, over eons, the typical, latter-day planning tools often cannot continue to meet current and future influences. A dynamic shift from traditional to progressive planning will normally have to occur. So what is progressive planning?

A. **Progressive planning is environmental.** It seeks direction from the changing world outside an organization for its basic forecast data. Research into changing trends and realities in the environment that have influence on the rate of progress in key areas of efficiency and productivity become a point of high priority. The models used to find solutions with research must be representative. Many use probability technology that falls short on scope and does not represent the total environment of the survey under consideration.

B. **Progressive planning is discriminatory.** Working capital would be selectively utilized on projects and activities that add wealth to the existing pool. Those decisions that give rights to actions that only circulate existing capital would be held as of lesser importance. Hence, they would demand less attention and less resources. Those are financing, manpower and operating structure.

C. **Progressive planning is courage.** Fear of failure is tolerated only if it adds motivational energy to the total manpower complement. It follows the general principle: "Better to fail because of misguided effort than to fail because of no effort at all." Lack of action, preserving old principles, is one of the major reasons why organizations fade away or die. Evolution dictates acts of constant fine-tuning and often major surgery.

D. **Progressive planning is creative change.** No new ideas are discarded without due diligence into their probabilities of success. Some may well be introduced into the mind or minds of humans with a talent to receive provocative, out-of-box ideas from the magna vault of the Universe. Many of the really popular and productive ideas were born in this way, that is, unplanned inspiration given a fair chance to engage its environment.

E. **Progressive planning is motivated people.** Personnel who believe in what they are contributing will always stay ahead of the curve of change.Without the energy of committed people driving the tactical structure with enthusiasm, the dynamics of success are low.

Readers Personal Space for Thoughts and Ideas

Micromanagement: Ego Power Trips under Phony Cloaks of Efficiency

To start an active debate or heated discussion, just challenge another human to define any proposition in minute detail. If the proposal seems to be instructional or merely opinionated, the mood of the discourse can reveal the basic attitude of each in the discussion. The Micro Manager of important projects can bring to view some very subtle motivations of those engaged in the process. Whatever pearls of wisdom rise to the surface as justification for the positions taken, the primary one most often thrown into the fray is efficiency. It is as if efficiency is the "ne plus ultra" of all possible reasons for a viewpoint strategy of micromanagement.

When the debate intensifies, the question may be raised as to why efficiency has become the bellwether of a largely psychological argument. Astute critics may see this stance as pragmatic nonsense tendered by its poser as a mere cover-up of the real reason why it is used.

The politician who wishes to put one over on the electorate will claim that his proposals are "what the public wants." With all the sanctimony he can provoke, he will place himself in the role of a god when, in fact, he may be very mindful that the public could be unaware of his pseudo role in the event. But it all sounds so genuine and demonstrably logical, so another boondoggle hits the street, in the form of micromanaged propaganda meant to deceive.

The fund-raiser who raises monies, caring only for his commissions, will promote, with a trick tear in his eye, all the soul-searing experiences he can find in his manual of operations, for example:

- Tear away at the guilt of his prospects
- Portray all the misery of the indigents all over the world
- Use needy little children to soften the hearts of the suckers
- Try to make all the public with their money ashamed of their own good fortune and order them to contribute right away so their billet in heaven will be assured with side orders of delightful rewards. That is micromanaged promotion that is meant to hook the last cent out of the pockets of cajoled prospects using vales of tears.

Donating is not a negative pursuit in itself. If done without lures based on human weaknesses, then it can fill a righteous place in the lives of true givers. However, if they are told it is a religious right to be made to pay, the true purpose leaves the leavening room in the lurch. The many skillfully constructed ploys to extract donated funds can be deeply resented by many of the so-called public who feel and believe they have a will of their own to make such judgment calls if the causes are truly genuine. Consequently, they may make a decision to avoid all donation requests, if once burned or twice shy from the wailing songs of sirens.

The telephone practitioners who have made it a business to artfully pick innocent pockets of cash have devised many tricks to maintain the interest of prospects. They glue prospects to the phone to hear of the "magnificent rewards they have won if they will only stay on to hear the good news." If they heed the siren song, they have been micromanaged into participating against their own common sense in many schemes that are fringing on illegality.

In commerce, the term "micromanagement" intends to explain the wasteful utilization of manpower resources. In large operations, for instance, where the credo of the organization is "get it done right," the lack of confidence by upper reaches of management in the forces on the floor creates a tendency to move executive energy downstream. In such cases, high cost energy from the top is misused at lower levels, where their efficiency is discounted. This method will create low levels of enthusiasm and, often, outright disrespect of the interventionists.

Persons on ego trips tend to micromanage their mandates in order to gain visibility and glamour, enhancing their influence. In the eyes of less power-driven individuals, these humans are pitied and disrespected. Taking credit or interfering with another's area of responsibility can create powerful feelings in those who have been thus abused, and retaliatory measures may occur.

Those occupations that give pseudo worth to effort tend to encourage tests in micromanagement. These include politics, advertising, marketing, designing, acting, public relations, and others where competition is fierce and general rules over ethical behavior are missing or loosely governed. When the mantra is known to be "all things said and done are okay, but don't get caught on the wrong side of truth," the role of management becomes a chaotic situation. This loss of good control of events is often excused under the blanket as creative restriction and will build a vacuum in the flow of productivity. It can kill the will of dreamers by putting them in cages, that is, the loss of creative energy through ignorance.

Micromanagement is not seen as being beneficial to its practitioners. What may be gained may also be stained. Reputations lost are a big-big cost in any organization.

Readers Personal Space for Thoughts and Ideas

(A) Renaissance Man: Sculpture of or Sculpted by the Iron Fists of Fate
(B) Rhythm and Symmetry: Two-Part Harmony in Verbal Intercourse
(C) Personality: Hot Wax and Honey on the Muffins of Communications

Renaissance Man (and Woman)

The term "renaissance" has its origins in earlier centuries when it was considered a renewal of interest in the arts and culture of the period. In more recent times, it has come to represent the talents of individuals who have shown special capabilities in many aspects of human behavior. Artistic and scientific accomplishments that affect a wide spectrum of the population are the usual measuring tools when recognizing a renaissance human being. The changing panorama in these fields over eras seems to have been mainly sketched by talents that have erupted in single minds and souls. Their contributions have upscaled public involvement over what might be termed the "thin veneer of civility."

However, if readings from the tomes of history accurately mirror the true nature of these architects of change, renaissance men have rarely been the macho personalities so often revered as examples of the "princes of history." They may present a façade of total confidence, but when they look inside themselves, they feel the insecurity that limits their close contact with the general public in communicating their internalized, masterminded ideas.

If communication skills are seconded to inventive genius, then it is logical that exposure of the products of renaissance minds to fields where the arenas of action are ingrained must be presented in the mix. New ideas are rarely accepted by the potential markets without a courageous, often very stubborn, paladin who partners with the creator and fights the negativity that usually will challenge any changes that affect a stable living environment. Agents, selling the production of artists, are a prime example.

Nor should it be forgotten that the flames of renaissance could burn within the ample female bosoms, uniquely framed and able to bear and bring all humankind into existence on this planet Earth. They need not cringe in a lesser role than those who don their dungarees one hairy leg at a time. As long as mother love prevails, female members will carry the flags of modern renaissance into the future as always. The role of females has been present since mankind has evolved and continues to be central to the continued existence of the human race.

With so many inventive humans, pushed by the impetus of evolution to create answers to changing needs, the seeds of renaissance are being planted every day in every way. Hope springs eternal in the human breast. Motivated forces are promoting new adventurers in creativity of all types. The future bodes well for the products of inventive minds. Thinking minds with courage.

Rhythm and Symmetry

The sound of good music stirs the human soul and sends shivers of reaction up and down the conscious body of its listeners. In many ways, the rhythmic cadence of a Viennese waltz, symmetrical and well played, rests eternally in the souls of those who are tuned to its rendition. It is almost a mantra to realize that great music touches deeper senses in the human body and mind than what one senses through the tympanic eardrums of the ordinary human. The rhythm and symmetry vibrates through other human sensors and can thrill the entire organism with a full appreciation that can be as specific as each unique and single life exposed to the sound.

The interesting factor is that both rhythm and symmetry are a component part of all types of good music. While one cultural group enjoys the sound of opera or symphonic music, it may have the same impact on others that tingle to the beat of hip-hop, rap, country and western, blues, religious, and sundry other genres of music generally. However, badly constructed music, constantly created and changed to exploit the innocence of younger humans for purely monetary reasons, will destroy their appreciation of good music. It has altered the impact of historic tempo, thus denigrating symmetry and reducing the sounds of modern music to loud, forced noise that intends to deafen listeners and overpower their sense of musical balance. The longevity of any one piece of such music and its immature, stagy presenters is short. Behind it all, the moneymakers, realizing the vulnerability of the genre, keep introducing new, young, attractive singers and boisterous, athletic combos to keep the flames of profitability well lit.

The world of classic poetry has suffered in its own way the loss of rhythm and symmetry. The low level of intelligent dialogue among a great many poetic aspirants has managed to make ordinary prose the bellwether of much modern poetry. True poetry has cadence, and the language resonates within the deeper sensibility of poetic aficionados. Even good prose usage has a rhythmic balance and carries a theme of understanding that goes well beyond the words themselves. Experienced readers recognize a well-constructed piece of writing and will seek out the work of good authors for the pure pleasure of sharing in their mastery.

> Fate hangs the golden gift of love.
> High up young trees of life
> Enshrouds it with the robes of joy,

Then peppers it with strife.
But later, caring, mature hands,
Well clad in gloves of worth,
Reach for that love so high above
And bring it down to earth.

Personality

Understanding the quality of personality leads one to experience the meaning of cosmic energy induced into certain humans to enhance their ability to project themselves and what they represent extremely well.

Personality is a subliminal quality with both positive and negative virtues. Because it affects its power by both sight and sound, it can be exploited for good and evil. In many cases, its possessor may not be aware of its presence but can notice that persons react rather favorably or negatively to its aura. This would support the thought that it resides in the subconscious spectrum of the human spirit and perhaps invokes its magnetic power as a part of the cosmic electromagnetism that encompasses the Earth. Because humans are a unique arrangement of cells within that force, not separate but an integral component, the flow of persuasive energy needs no exterior transport but is likely just a small transfer within the total human entity.

Natural personality can be neither bought nor sold. Why some people, but not others, have such a persona is not generally known. Is it an inherent virtue that has strayed willy-nilly into one gene stream but not another? Or is it more an image than a reality, less for what it is but more for what it seems to be? Or could it be that some humans have a capacity to imbue their subconscious selves with some cosmic power from the universe and thereby gain an aura of magnetic influence that attracts the conscious awareness of others? The personality of those who are prone to exploit wealth and power is seen as surrogate straw. Personality, unlike mists on the Matterhorn, never hides the real virtuous grit of its fortunate possessors. It can be neither envied by its cravers nor denied by its holders.

To be blessed with the human quality of personality is a gift that has value far above it visible and subliminal virtues. This power of attractiveness can be exploited by those who choose to sell the benefits of the quality for no more than the mighty dollar. In doing so, they may well incur the negative energy that it can generate. That negative force, if not corrected by the possessor of personality, will gradually isolate the person and drive them to compensate with possessions that have no real traction in tactics to buy companionship and affection.

Personalty is a quality of merit on the plus side of life. However, the astute possessor will realize that its overt exploitation can raise envy, disgust, hatred and other negative reactions if the very possession is not thoughtfully and congenially shared.

Readers Personal Space for Thoughts and Ideas

Poetic Interlude: Take a break from Prose

The human spirit writhes and curls
To find its proper place
On tortured tracks of destiny
Controlling time and space.

<+><+>

But human eyes see only that
Which suits their narrow view
Of what this life portends for them,
The false as well the true.

<+><+>

The pessimists can only see
None but a perverse face,
And optimists, they tend to scan
The docile in each case.

<+><+>

But life, that mystery for all,
When lived from end to end,
Commends each day to find a way,
To let it best extend.

Paradox

There are no words on Earth to say

Why one of us is whisked away,

Or should we think it is because

We're needed for a higher cause?

One thinks that Fates they set us free

To serve in God's eternity.

The Wish

Through wisdom, may it be our fate

To stumble up to heaven's gate.

With knuckles bared, beat a tattoo,

And pray compassion gets us through.

St. Peter, May I make this clear,

"You'll save me from the devil's spear."

Readers Personal Space for Thoughts and Ideas

The Magi in Imagination: The Thaumaturgic Carpet in the Human Mind

Like all things in human experience, because something is part of our everyday lives, we tend to take it for granted and ignore its relevance. Nothing brings that home better than our treatment of the magical tool we call our imagination. If we take the time to analyze it, how many times in one day do we resort to the use of this facet in our mind? Perhaps a hundred? Or more?

Imagination is inborn in every human being. It is part of the blueprint of existence and works in concert with the instinct of survival. It is an independent force in the human mind and is like a kaleidoscope in making many ordinary acts, ideas, and events both colorful and mercurial. It can paint a picture, take a trip, play music, and be the virtual exponent of all activity that can be desired without requiring the bodily presence on location.

Without imagination, it is highly unlikely that humankind would ever have managed to evolve from simplistic origins to the highly sophisticated creature that now inhabits all corners of the globe. Its impact on the changing conditions on this planet begs the question. What in the human psyche drives the need to innovate and develop improvement of the existing status or create new instruments of change? The usual answer centers on the belief that the mind of humans is supreme. This attitude, an ego-focused obsession, finds no proof to suggest that there is a relationship between the mind of humankind and the huge sheath of energy that dominates the Universe. Still, one can wonder how it is possible that no formal connection exists in the communication that may occur in that multilateral environment that exists in common. It would be unlikely insanity to think that the common energy does not relate with earthlings in some way.

The fact that so many of the eye-popping inventions that we experience in our modern times did not exist in ages past may give credence to the thought that the human mind has gradually matured over time. With advanced wisdom and a more courageous insight, continuing evolution of imagination's creativity, and a willingness by humans to accept the influence of the cosmos, a world with a mysteriously enriched environment, is almost a foregone conclusion. The pressure of evolution will surely influence the changes that will color the history of the future for Homo sapiens on our Planet Earth. On that we can believe.

Where will the imagineers who can change the scope and breadth of those ideas that change the human environment, come from? The planet Earth is rife with raw and vibrant talent in the creative segments of society, but most do not have the courage to face the negative forces that abound. These forces seem to have as their mandate, their ill-chosen need to keep the talented ones from raising their image above the massive sea of mediocrity.

Which raises the following question: Is it possible that the rules of equality are such that a balanced populace must prevail in society in order that conflicts can be annulled or mediated to bring peace to the concept of community?

It is a known fact that, in a democratic structure, there is a natural presumption that no unconventional, imaginative event or construct be accepted without critical and often envious challenge of its right to exist separately. Is this then a universal force that tends to even out the anomalies, created by the talented few, for the benefit of the turgid masses? Have individuals lost the right to their personal convictions and ideals, and does fear of consequences rule supreme?

In thinking about the role of universal impact on human rules, in matters that exist to ensure that a smooth transition occur in evolution, the pressure for reducing the crags and spires of unbridled creativity do face natural tendencies to find some level of equality. This tendency is seen in many instances of human endeavors. The idolaters of the horse and its role in mobility severely criticized the Ford phenomenon with automobiles, a creation that was destined to change the face of transportation in the world. The imagineers who thought that humans should be able to fly like birds brought aircraft onto the lists, against the naysayers who saw these ideas as ludicrous. That major shift in thoughts and feelings has changed the concept of community isolation to one of a planetoid community around the globe. The innovative, imaginative design of printing by machines and mass media left the idea that smoke signals and writing would be destined to occupy the popular convictions of the masses long left behind. Those too have suffered the fate of obsolescence when the cosmic energy of the universe was found to have qualities heretofore unknown in the magical field of communications through outer space.

All the major changes in the environment of planet Earth were occasioned by the power of imagineers. Those are the humans with the minds and souls of courage to defy the challenge of a status quo and seek new dimensions in the evolution of humankind on planet Earth and the Universe itself.

Perhaps in time, the concepts of imagineering will gain traction, and the growth of mass imaginations can continue to evolve still further to enrich the lives of all Homo sapiens as the future of our destinies expands. Expansion of the imaginative mind needs challenges and those are raised by authors, designers and thinkers with the urge to lay it out and make it happen.

Sic temper fidelis.

Readers Personal Space for Thoughts and Ideas

Pure Logic: Bedrock under Life's Realities

Every human, past and present, has faced or is facing the realities of life. Nature has set rules that cannot be denied. The main ones are the urge to reproduce and the inevitability of death. One does offer some flexibility in terms of choice, but the other offers no loophole of escape. Therefore, we must believe that the only true reality that signposts the stages of our existence on this planet is the last and final one. The gap twixt birth and death is minute in cosmic terms.

Whatever choices we encounter on our pathways of destiny, none have such a clear endpoint as that of the last breath drawn. With that ever-present saber of Fate hanging by a hair above our heads, one would think more people would take their state of existence more seriously. Pure logic should define the necessity of using the relatively short span between life and death productively and sensibly. Why then do so many humans choose to defy logicality and fritter away, so callously, the only opportunity they may have to build something worthwhile? Worth some deeper thought? For all humans, of course.

Unfortunately, logic is not a universal persuader. It might be better considered an ultimate thought weapon, conveniently used to support a point of view in heated type A arguments. Because it is a personal belief, its interpretation is only valid for each protagonist. Used as a lever to convert non-believers into believers, it can become a brainwasher that might convert the minds of vulnerable humans to become primary reactors rather than logical evaluators; ignoring the human right to personal beliefs.

There is common logic in most situations where its virtues are accepted as rules. In issues that require a higher quotient of integrity, the use of pure logic to clarify the attitudes of those involved cannot be decried. In those cases where life-and-death situations may be involved, attempts to find the truth behind the proffered facts does engage logical analysis to prove the rule. With issues where contractual dialogue is a vitally significant component, the same principles can apply. However, in facing reality, not all situations in life will pass the test.

Because logic is a philosophical concept, it is not generally suited to the daily thinking of earthly humans. Like so many esoteric words with meanings that float in space, logic becomes a tool in some situations but, because of its missing boundaries and universal definition, is used on rare occasions in the world of common usage. Perhaps the theoretical idea of logic is understood better as its usage in everyday support during normal dialogue.

In the field of mathematics, logic is a defining element. It can enter into proving theories and supporting formulae. This type of usage, however, is so theoretical in most instances that it can be applied only directly by those experienced in solving problems with number-manipulative techniques and specialized dialogues. Hence, its usage cannot be generally flexible.

In conceptualizing the scope and functions of the Universe, however, it can extol the idealization of many suggestions about the evolution and probabilities of one's cosmic reality. Because human designs for such speculations are biased on logic that emerges from human minds, the images of cosmic activity and likely consequences are as painted by human imagination. The application of logical boundaries is therefore somewhat conjectural and gives traction only to the subjective enthusiasms of the proponents. Because proof of cosmic reality cannot yet be offered from any point of cosmic fact, but only from the perspectives of earth scientists, such prorations may need be spiced with the proverbial grain of salt.

Because logic is an original idea by and from earth-based science, it is a useful ideology when used here on planet Earth for settlement of issues and arguments. It ceases to have purpose in the endless speculations that occur when trying to define the mysteries of the Universe. Perhaps as humankind matures into spatial realities of new and expanded dimensions, it will begin to find a place in human philosophical thought and action about what is occurring out there in cosmic space.

Reader's Personal Space for Thoughts, Ideas, and Arguments.

Feelings and Attitudes:
Prime Keystones in the Arches of Life

From a panoramic slate of choices, these keystones are able to cover all the emotional facets of human communications. Both are necessary to prepare the stages for interaction on most levels when people call or are called upon to mingle with their fellows in the ordinary course of living. Feelings are intimate and private to all humans and are part of nature's inborn defenses to protect the soul.

Babies, from the moment they are introduced to the world, express their feelings with their voluble voices. Their hunger, fears, general unhappiness, need for comfort and affection, and discomfort must be tended to if the newborn is to survive into maturity.

Feelings can be all the colors of the rainbow. Logical or not, pleasant or offensive, raging or benign, or considerate or uncaring, they are the defining forces that decide what every single human will become in his or her lifetime. They will limn the criminal and extol the hero. There is no escape from the talons of feelings.

Because they are composed of both positive and negative emotions, it becomes a challenge in life to learn to control them with the power of one's will. Forces in the religious community, realms of politics, educational institutions, and experiential situations can and will be there to help a person to become a useful element in society. The fact that these forces may be self-serving is a moot point, but they can be teachers in the interpretation between good and evil.

It is unfortunate that the basic foundations of positive development often fall short of maintaining earlier lessons of exemplary behavior. The power to resist the siren songs of purveyors of evil later on in life depends on early teachings. These usually commence in the parental home where examples of good and bad habits are exemplified on a daily basis. Much of youthful learning is begotten in the very early years. The importance of the home environment is unmatched by any other teaching process. Adult insecurity is born right here in the cribs of human maturation.

Much of the youthful indiscretion that prevails in open social systems can be traced back to the poor grounding of evildoers in their early years of development. Etched into early minds by the bad teaching of their parental mentors will distort the rules of right and wrong in immature minds perhaps forever. From this are born the most basic rules of living and the attitudes that govern all aspects of life itself.

Attitudes are the human energies that deploy the lessons learned from the wide bases of feelings. While feelings are the introspective forces, attitude conveys to the exposed audiences just what the human behind the attitude is really made of. If the emotions from feelings coating the attitudes are not perfectly aligned, the first will affect the desired impact of the second in an unforeseen manner. The tones of attitudes will generally reflect the emotional feelings that color them. A hostile feeling in a relationship will ensure that the attitude of communication between the parties will be negative and unproductive.

If attitude is the view that observers note of anyone outside of themselves, it behooves those who wish to portray a positive personality or character, to ensure that their attitude will truly reflect the emotional feelings behind the view. An example of the energy that can project a message into the minds of a mass audience, using feelings to propel an attitude of conviction, was paramount in the hysteria of the public that endorsed the commencement of World War II.

A feeling based on revenge, seconded by an attitude of supremacy forcefully presented, defeated common logic, and a killing chaos prevailed.

To present the other side of the coin, one might think of the cause of Jesus Christ, a humble man filled with feelings of love and service to his fellow man. He was provoked and reinforced by an attitude that portrayed him as a messenger from God, saving the souls of humans in the world. His dedication and stern belief in what his calling in life was intended to be created Christianity. The fact that this attitude has prevailed is proof of its validity.

Since the rules of Christianity were penned by human minds claiming some kind of divine intervention, the feelings and attitudes of religious pantheons must be carefully weighed to iron out the speculations and biases of the writers.

Feelings and Attitudes. Governing the inter-action of humans and therefore a critical set of emotional influence. They must never be discarded by the progress in evolutionary change.

Readers Personal Space for Thoughts and Ideas

Mental Energizers: Prospecting in the Mind

In the world of English language, some expressions are repeated so often that they become clichés. It is necessary, therefore, to create new ones from time to time to keep usage in the dialogue of everyday communications and add some spice to the nattering verbiage.

Some languages are richly evolving over short periods of time. Others are expressive and colorful. Then there are those that are romantic and those that are academic. Scientific and poetic, didactic and entertaining, the purposes defined to communicate as succinctly as possible to general or specific audiences. The ones that pose challenges are often those that grant mental satisfaction. Some examples can set the stage:

- Getting a gift unexpected is neat.
  - Giving that gift is excruciatingly sweet.
- Toast 'n jam with a kiss every morning is great.
  - Traffic jams late on Friday breeds curses and hate.
- Cussword puzzles are not acrostic dilemmas.
- Honeycombs don't part hair very well.
- If you lust for wild rides, off road and off rail,

- o Grab a hold of a Bengalese tiger's long tail.
- If your house is on fire and you're there abed,
  - o Don't phone for an angel. Try Satan instead.
- In 'roo land, "g'day" is hello and good-bye.
  - o Aussie ladies like quickies. Perhaps that is why?
- Bathroom mirrors, like lions, have a mean tooth.
  - o Hangover mornings are a hard-bitten truth.

Readers Personal Space for Thoughts and Ideas

Equilibrium: Impossible Balance of Forces

In the theoretical minds of humankind, the myth of a harmonious existence between cultures that can be enjoyed by all universally is espoused and idealized as an ideal dream. The cosmic rule of equilibrium is that it is a force of nature that is believed to be one leveler of emotional discord, not only among persons, but also among nations, cultures and all living sentient creatures.

Humans, whether they recognize it openly or not, do strive to engage in deeds that attempt to invoke equilibrium as the answer to restore a form of balance when events or situations go awry. The reason for this would appear to be a desire to return to a state of known and experienced normality after periods of dysfunction end.

The state of equilibrium is also an established position with governments, commercial operators, social organizations, and almost all major decision makers in life. The governing factors are usually established so a modicum of real control can be set and monitored. Without a state of equilibrium, chaos would replace control with all the negative factors that that status would entail.

In the matter of control, this is supremely evident in situations where the profit motive prevails. To ensure that future monetary results are to be positive, budgets and strategic planning define the path along which manpower energies are directed and audited. The concepts of equilibrium set the limits on the balance between revenue and expenditures. When one or the other fails the established objectives, the loss of equilibrium may indicate an early warning of some difficulty, after which, corrective measures can be applied.

Historical evidence may be used to understand the reasons why equilibrium is facing extinction as a force of control in the lives of the peoples of planet Earth.

Sometime in ages past, the family unit was the bastion on which a life or lives, based on equilibrium, could serve to support and foster sensible development between those of the human race. The utilization and optimization of human energy was allotted according to the available resources. The fathers were the muscled ones, so, in the interests of survival, they did the heavy lifting in a functional family. The women, married in as partners, did the work that utilized their minds to plan and execute the administrative and living standards for all members of the working team. The children, viewed as energy sources, generally fulfilled the roles of supplementary bodies used to do both light and heavy work as their maturity allowed. To be a part of a functional family, all had to toil in the interests of surviving the rigors of mere existence.

Such early families worked on farms and forests. The food they needed was mostly home grown, and while it might have been sparse and simple, it did supply the amount of energy required to perform the duties of a rural economy. Equilibrium between the need and supply was rigorously maintained if progress was to be assured.

When the supply of energy grew beyond pure need for survival, a sense of imbalance caused the development of urban resources to restore normality. The urban communities had their own functional equilibriums to maintain, and this hosted multiple changes in the environments, both urban and rural. In time, because machine and equipment energy could replace and multiply the use of human energy, the traditional surplus developers sought new venues of equilibrium. The extensive growth of energy freed up the pressure needed to create activities that would ensure a better survival. The growth of a leisure industry, creative arts, and sciences intensified knowledge bases, exploding and creating different levels of disequilibrium that tended to destroy the traditional peace and comfort of the masses. The modern age has been born out of that imbalance, and continues to feed the evolutionary pace of change.

With excessive human and mechanical energy available, it has quickly surpassed the equilibrium center of balance, and that has invoked a tremendous shift to creative development. That phenomenon is evident in the support of investments in transportation, communication technology, administrative tools, agriculture, forestry, housing, commercial and residential architecture, construction technology, and improvements in all phases of a modern's living standards. Adventuring in cosmic space has ingested huge quantities of excess energy and resources, and the search for a modest equilibrium has vanished in the vanguard of created chaos and dying dreams.

Industry and commerce has been forced to seek some answers by engaging a philosophical attraction to the concept of "big is better." In this manner of maximizing the use of available energy, huge operating plants, towering residential structures, massive box stores, and extensive malls have dominated the environments and have virtually decimated the many small historical units. In so doing, the populace has been forced to accept a state of disequilibrium and to find new ways of coping with a strange and often forbidding challenge of survival on a planet that spins with attempts to neutralize chaos.

Like in all excessive ventures of humankind, cosmic law will tolerate dysfunction only for so long before ameliorative influences are imposed. In ways that only unseen cosmic wardens can apply, incredible changes in the natural environmental parameters have been invoked. These are best experienced in hurricanes, typhoons, earthquakes, pandemics, sterility, and other yet-to-come challenges that have been called "black swan" occurrences. All nature's events are not preventable by humans as they are part of cosmic energy directives. The drive by nature to restore equilibrium is unassailable, and we, as Homo sapiens, can only observe and suffer the consequences of our avaricious and materialistic devolution of the balance of equated forces. Perhaps a second eviction from this glorious Garden of Eden is forming?

The future of humankind, primary custodians of the perimeters of planet Earth, is dark. Still clinging to the truths of history, most are reticent to grasp radical out-of-the-box ideas for the return to equilibrium. In a wild conjectural mood, one might think that we, of this universe, are preparing the conditions for another big bang explosion.

Readers Personal Space for Thoughts and Ideas

Conscience: Nature's Punching Bag in the Human Comfort Zone

For those who claim to have no conscience, there is an eternal invitation to join the dissimulators club at its headquarters in hell. No matter where humans try to hide it, it is one of nature's unassailable tools aimed at keeping some equilibrium in people's aspirations and other general misdemeanors.

A primary dream in life is to settle into an emotional zone of personal contentment where all aspirations have been attained and where one's emotional tides no longer need to flow to and fro catastrophically. Very few humans ever reach that Valhalla in one generation, and the guardian at the gate is none other than an unseen but never unfelt inner voice called conscience. Born in the womb, it acts with a stern and generally objective voice when duty calls. To try to limn its purpose in human lives, it often appears in the mind when a human attempts to make offensive moves on a fellow neighbor. Conscience, without fear or favor, rules with the mandate that, for every decision that is made, there is always a boon or bane to be undergone. Positive virtues carry negative tattoos into the fray and the opposite can likewise prevail.

With excessive greed as its opponent, conscience must face that ferocity with logic and a mind-set that is soaked with the ability to define the difference between right and wrong. Guilt, shame, fear, insecurity, and bodily indisposition are in the basket of tools used when an unworthy situation or event occurs. Such excesses, some of which can be defined, are extreme greed, jealousy, hatred, lust, prejudice, sloth, dilettantism, covetousness, and supreme power, along with a host of other feelings that bolster human nature in finding ways to dominate the minds and bodies of their fellow humans. While even in the throes of these excesses, one might entertain some compassion for the lot of such compatriots. However, lest conscience takes over, nothing will intrude to effect the necessary changes in attitudes. Apologies for the devolution of conscience are many and easy to embrace. More's the pity. Live and let live sets up the rules.

However, recognizing that conscience is a cosmic element in life, it must remain as a leavening force in the souls of humans although it may continue as a tattered reminder of its higher virtues in the history of humankind.

To understand conscience, one must start with a critical inward look and from that subjective landscape, separate the positive ideals from the negatives. Once that is done, the sense of conscience can begin to gain a foothold in helping to understand the human soul.

Readers Personal Space for Thoughts and Ideas

Humor: Champagne Bubbles in the Lager of Life

Defining humor is like trying to bag fog. It is a human sensation that is both natural and created and therefore obeys no standard laws of understanding. It can be inborn for some and totally created by others. It is used to leaven social interaction, to amuse the masses, and to establish rapport in many situations. It sounds like a miracle drug, which, in truth, just might be its final judgment.

Humor is said, among its many virtues, to be healthful. A good laugh a day is said to extend one's existence on this planet in some truly measurable manner. A sense of humor is straight out of the toolbox of nature and can use infective devices to stir the soul and excite the human spirit. These are many, but to limn a few, take note:

- **Situational:** The amusing antics by those who earn their living from the exploitation of humor are legion. The minor entertainment industry with stand-up comics and television shows are a time-tested rage. The stab of humor might be slapstick, pie-in-the-face ribaldry, or the exploration of a personal fall from grace, a created situation to satisfy the human urge to see pleasure in the trials and tribulations of one's fellow wretches.

- **Incongruity:** In a general society, there are recognized, long-term norms of conduct. Testing incongruous interpretations of these norms with out-of-shape dialogue and behavior, often sick or silly, evokes hilarity among those observers who do get the joke.

- **Communicative:** Where there is a need to titillate the buying urges of consumers, humor plays a vital role. It may irritate or make viewers feel good, but it plays a sterling role in such promotions. But if the satire is ill placed, it can be disastrous. Lady Montague once said, "Satire should, like a polished razor keen ... Wound with a touch scarcely felt or seen." A lesson for those with a prurient colloquy and a thin sense of temperance. Humor has a built-in cost. As most humorists know, it is a blade that cuts both ways.

Readers Personal Space for Thoughts and Ideas

Infidelity: Mainlining the Life Sport Locomotive

Can this be defined as one of nature's gambits or a technique invented by man to ensure that the rewards under nature's laws are fully enjoyed throughout the short interval of one's life span? The fact of infidelity, once thought of as a male failing, has segued to both sexes with equal emphasis in the last fifty years or so. Unhappy with their roles as the aggrieved spouse in what was intended as a state of equal opportunity, the female has come much closer to equality in the new age. With two major developments, the female partner has arrived at parity. These are:

- With the arrival of the Pill and other modern methods of reliable contraception, fear of pregnancy is no longer a factor. Free of consequences, females can aspire to the role of aggressor in the arena of a sexual opportunity. That was a delicious victory in the battle for equality. Female humans can now express their deepest feelings about their sexuality without fear of recriminations.

- Pressured by the rising costs of living and a rewarding lifestyle, many wives have undertaken double roles.

Not only are they still bound by maternal factors to be keepers of the household, that is, having children and maintaining the family circle, but they have also, out of choice or necessity, become full partners in the responsibility for earning incomes sufficient to the needs. This has also added to the cause of female independence, which has thrown the shibboleth of infidelity into the cocked hat of new rules for both sexes.

The rising power of the female partner in a mutual relationship has not been met with total enthusiasm by hide-bound males, who now must acknowledge that sauce for the gander has become sauce for the goose as well. Recognizing that it has opened the door to female infidelity on an equal basis has brought new male concerns about sex and marriage into focus.

Infidelity, once a tasty treat for males, feels like cotton in the mouth when weighed in the equality equation. It has provoked a reticence for marriage among male prospects and a lifestyle that no longer seeks the seals of a marriage contract. It proposes what is now called "friendship with privileges." While such unions have gained traction in the new age, the problem of dealing with children from such couplings is now the new varmint in the woodshed. With nature's call for motherhood as strong as ever, if not stronger because of the new circumstances, new age women may still have to pay a price for their freedom and equality. An early trend to changing back to the family of equilibrium, so long a former tradition, has been seen creeping into the decision cycles of the sexes.

The minds of humankind are not alone in the decisions that must be made in the future. With Mother Nature pushing harder than ever to guarantee the continued existence of the human race, she has a hand in the game. In the long term, the cosmic rules of propagation will prevail, and the human race must find a better formula for ensuring the sexes can abide with each other in a form of mutuality. Whether that can contain the rules of true equality will be proved out by the flow of history as humankind matures into the next stages of life on planet Earth.

To paraphrase Omar Khayyam, "The hand of fate writes on … And having writ … No means of man can e'er erase one single word of it." Let us bear that in mind. An ideology exposed is a rock that lasts through thick and thin for a long time.

Because the urge to merge is a law of nature, the consequences of moral laxity will continue to plague the human race unless a better state for both sexes can evolve. The current rules are fraught with subterfuge and immorality that tends to pressure the margins of both wedded and unwedded bliss to the point of desperate frustration. While the attitudes of jealousy continue to test the depth of conscience, there seems to be no easy answer to the case of infidelity. Not a happy state.

The most telling factor in the search for better answers to the infidelity dilemma will, in full conscience, settle firmly on the price that is being paid by the children born out of experiments with infidelity. The irregular home environment, so much a telling component in the role of parenthood, has left a host of dysfunctional offsprings in the lurch. Without the stability of committed relationships, there is no way to find merit in the function of good parenting where loosey-goosey, sex-for-fun-only is the prevailing modus operandi. Deeper thinking of the consequences and better rules for managing birthing privilege are up for grabs.

The opportunity to reverse the damage is in everyone's mind field.

Readers Personal Space for Thoughts and Ideas

Time and Space Realities:
Some Rules on Planet Earth

The obsession of humankind with time and space considerations is a known reality. Without laws to govern human occupation of earthly space and rules that govern utilization of and restraints on time, the laws of planet Earth would be chaotic.

To compare these rules and laws with those being paramount in studies of the Universe is axiomatic and very difficult to grasp. In the void of cosmic space, there are no limitations on space as it is infinite and expands or contracts according to the demands for universal events. In earthly terms, cosmic space has no relevance to humans in their considerations of space allocation and control.

Again, in cosmic time, because it is eternal with no beginning and no end, it fails earthly logic where every event must have a start and finish. The time utilized is measured and, in most cases, is costed out in considering its value for specific applications.

Time and space on planet Earth are every bit as eternal and infinite as they are in the Universe. The factors that change their utilization on earth are human-made rules that have parceled out the space into continents, countries, lesser tracts, and right down to the personal space that every human in the world occupies.

The necessity to measure time in units by eons, eras, centuries, years, months, weeks, days, hours, minutes, and seconds was due to the defining of events and pricing of time. In fact, this measurement was created to evaluate the utilization of energy, both human and mechanical, in the process of distributing available resources. As humans expanded their presence in the world environments and the use of monies to value energy and time equitably gained legal stature, the system has evolved with both advantages and disadvantages over time.

The only true space that any one human possesses is that which is contained within his or her skin. The planet Earth owns all other space, as it cannot be easily shifted off the face into outer space. The apportioned space of deeded or leased property, while it is given as a human right, is basically only on loan from the planet during the lifetime of any so-called "owner or lessee", using a piece of paper to give permission to occupy and utilize a piece or pieces of planet Earth. If all the humans were suddenly to vanish from the world, all property would automatically return to the planetary inventory.

In the rules of the Universe, all matter therein contained is not parceled to planetoid owners but is forever granted as a cosmic right to develop and maintain for the betterment of its occupants. In the planet environment, however, property ownership and utilization is a legal right, the laws for which are created by and for the benefit of privileged human possessors who then mete it out to the publics for a monetary consideration and for specific periods of time. Under common logic and cosmic law, this process could be called human exploitation.

So it may be assumed that cosmic time and space in the universe, and therefore in all its component parts, exists to fulfill the demands of evolution and the creation of new sentient life throughout. If humans forfeit common logic and abort the cosmic rules that govern use of resources, some corrections to the methodologies of exploiters, by the cosmic force, may prevail. This correction could be by changes in the topographic environment of earth, reduction in the human population by various means, or drastic extermination of planet Earth in total. All such measures are well within the power of the universe to put into effect. Armageddon!

Such drastic remedial events would not happen overnight. The changes take millennia in human time, so long periods of grace may be expected before any ultimate devastation might occur. This means millions, if not billions, of earth years in the future.

Measurable time and space concepts are human creations, and making these factors important outside the planet Earth environment is a conjecture in any case and will continue as such until proof of cosmic realities can be guaranteed.

However, unless severe trials of conscience are to become a public norm here on planet Earth, the reckless exploitation of resources for the benefit of some classes of its occupants at the expense of better husbandry, the time frame for existence of the sentient creatures may be shorter than expected. It should be carefully considered that the planet Earth exists for all its flora and fauna and they have the right to insist on a fair and equitable of use of all natural resources.

Readers Personal Space for Thoughts and Ideas

Taps : The Last Post in the Macroscopic Series

The experiences and conclusions in the Macroscopic series were obtained over 80 years of practical research in various parts of the United States, Canada, the UK, South America, Japan, the Baltics, Italy, Switzerland, France, Denmark, Sweden, Finland, Norway and Jamaica. The focus of the research was to understand, not what was happening to humans, but why it was occurring. Contacts with over 2,000 humans defined the basic observational findings and set the stage for this series of theoretical, philosophical and metaphysical musings. This series is only a small selective part of the total findings but if they challenge and inspire readers to expand and think further on the various issues on their own, then the objectives of the book will have been fulfilled.

Exx-Rays from the Aging Cage (Prologue)

The blueprint for all living organisms on this planet Earth outlines the basic pattern that must be traced on the pathway of existence from conception to demise. All forms of life, whether animal, sylvan, avian, piscine, insecta, hexapodal, or microorganic, are subject to the same laws of existence. They begin with conception, move forward to creation, and then go through to growth, maturity, and productivity. Then there is aging declination to the final stage of dying. There are no options with the defined pathways of existence. Cosmic laws are based on evolution of all things through their cycles until the expiring constituents revert to their origins and support the emergence of the next new generations.

One of the major phases in the cycle of human life is the period of aging. Its gradual emergence as a fact sets up new, though gradual, sets of circumstances that we humans must learn to abide by, embrace, and sustain. Some of the challenges of that period are what are defined here as coping in and out of the aging cage.

The Aging Cage is not a structure made of materials but is rather a state of mind. Each of us will know it through the aging period until we stand on the threshold of our own last flight with a boarding pass in hand.

Senior humans create the aging cages, and they will take their shape, subtly chosen or forced by circumstance to be occupied. They may have many doors or none at all. That depends on how the omnipresence of demise is faced and handled. It is up to every person based on his or her historical experiences, feelings and attitudes.

Like all periods in the cycles of existence, aging has it joys and tribulations. But it is an exemplary period and should be embraced for the benefits and tranquility it bestows.

Exx-Ray Visions from the Aging Cage Contents

Cosseting the Super Seniors

It is controversial to suggest that senior citizens of planet Earth are in some sort of aging cage. However, the definition of this cage is a mental structure from which one cannot escape. What better definition for the natural state of growing old. It is a phase of life from which, at present at least, there is no way to live beyond one's personal blueprint of existence. Hence, it is an attitudinal cage. The length of tenure is, however, different from one person to another.

In this case, the cage is a mental vision, and in contrast with a real metal cage that has but one door, the aging cage can have many doors that permit exit and ingress at the will of its human possessor. Those moves, both in and out, are what this set of writings is all about. It is possible to conceive of situations in one's dotage where the imagination can find the key to spring the latch. It is the challenge. Yes, all those whose history can measure the years of seventy-five or more, super seniors, are the ones who have won the lottery of the years that have passed and see the specter of the golden door somewhere down the road ahead. Experiential exx-rays have explored, with inmates, exposure to the positive and negative challenges that have been faced and will be faced by ongoing aging minds and bodies as each life evolves.

The onset of aging symptoms can vary from person to person and culture to culture. Much of what happens as the years pile on can be genetic, upbringing, experiential, attitudinal, or fortuitous. In many ways, the game of life is one of chance and uncontrollable. However, if one chooses to abuse the body during excessive youthful gambols in a world of tobacco, alcohol, drugs, and physically harmful exploits, the payback may inevitably be a very short stay in the Aging Cage. Poor health in early and midlife portions of one's destiny spells extended misery in later years. That means being confined to a doorless cage with no way out. The virtues of some asceticism and topical physical exercise during one's productive lifetime cannot be overemphasized as insurance against early senescence and unhappy years suffered later in unforgiving misery. It is a somewhat impossible dream to paint a picture of a rusting hutch for vibrant youth in the throes of imbibing all the elixirs of a worldly smorgasbord, catering to the vulnerable desires of youth. Unwilling to minimize the damage. Blind to the consequences that may lie far ahead. These obsessions with youthful fun ingredients can be lethal.

The incidents of incurable health problems, such as Alzheimer's and Parkinson's, arriving uninvited at the early stages of aging, are, at present, challenges of fate that are akin to symbols of the law, ready to incarcerate those who are quite unprepared. Therefore leading them into their aging cage and perhaps welding shut the door.

The cost and consequences that color the attitudes of aging seniors are herein described from the diary of the author. One cannot feel the emotional psychology of aging unless one is in or is getting there. After ninety years of coping with the demands of early survival, health issues, schooling challenges, wartime experiences, academic struggles, and missed life's opportunities, the items that are scribed here draw worthwhile validation from the pages of reality. Histories from some 200 World War II veterans have added color to the research.

Lessons learned and other's life adventures isolate the main components of rich experiences in the aging cage. They are briefly detailed as follows:

A. **Assets:** A comfortable couch in the aging cage can only be assured if sufficient monetary resources to cover the expected costs, support the years in retirement. Stress-free living conditions, whether in residence-based or retirement homes and health-care facilities, demand financial funding in excess of any government pension support. The choice of lodgings in senior years should be well considered during one's effective earning years. Savings and investment income should be programmed well in advance to ensure a pleasant interlude in the fading years of one's destiny. If blessed with good health into one's eighties, part-time employment of a productive, yet stimulating type can add income to a discretionary pot without strain. Charitable effort is a use of energy where cash is immaterial and excess time prevails, but it does supply topical bonhomie when it is needed the most.

B. **Attitude:** Of key significance, when occupying the aging cage, is attitude. It is a state of mind that is self managed. Negative vibes will attract negative reactions. A positive outlook in this phase of aging will make life well worthwhile as it will ease the transition from the controlled pressure of business life to one where all the control is contained within oneself. Making that transition without losing optimism is a must. It is a problem that many retirees have deep psychological problems with. Letting go of the tensions of career employment is never easy but is necessary for a welcome stay within the aging cage.

C. **Exercise:** The aging bones and muscles that should be endemic to a good life will decline progressively and persistently in the aging cage. The will to engage in lightly rigorous and consistent exercise of both mind and body is one of the most difficult tasks to consider. The mere idea goes contrary to the dreams of relaxing thoughts and idle hands that colored the mind during the sweat and pressure of a career. And yet, without exercise, the mind and physical muscles of the body will gain debilitative momentum if left to loll in idleness for extended periods of time. Of course, the challenge of marathons, except for supernaturals, will no longer be possible. But many sports such as tennis, golf, walking, bowling, skiing, and gymnastics can and are adaptable to the needs of dedicated cagers on the prowl for tactical activity. The aging schedules should be shorter for physical effort and longer on the mental side. Exercise for the mind is strategically mandatory. The natural loss of short and long-term memory is an experience no aging human can enjoy. It imposes stress that can affect self-confidence, sociability, effective communications, and daily behavior. Loss of mind/ muscle coordination can create a hostility within oneself that is anger at the carelessness that occurs at times without rhyme or reason. While the overall fading of the mind cannot be stopped, it can be slowed down with designated exercise. Daily exposure to acrostic crossword puzzles will exercise the mental cells in useful ways. Jigsaw puzzles also stimulate the brain. Model building, along with technical and musical training, develop mental strength and often physical agility. Artwork, painting, and writing offer challenges that can be reasonably difficult but interesting and are mind-bending challenges. Reading keeps the ennui of boredom at bay. The idea is to keep both mind and body engaged. Only then can the loss of neurons be decelerated and the creation of new neurons provoked. To produce new neurons seems to need tough mental challenges on a consistent basis.

With all the cylinders of mind and body well occupied, the time in and out of the aging cage can be a most stimulating experience. And that is the main benefit to be enjoyed during those footsteps down the dusty trails of destiny. The relief from stress itself can be a bracing experience. If one can experience all the good that can come with aging, there will be no reason to abandon a comfortable aging cage. After all, it can mitigate the pain of coping with negative problems and build on the satisfactions that should and could be enjoyed.

Readers Personal Space for Thoughts and Ideas

Keeping All the Marbles in the Bin

As we probe the atrophying lump of tissue that is normally called the brain, the transitional stages from full utility to semi-modality is faced with both glee and terror. Nothing can be as frightening as losing the brain's capability to function effectively. Testing the status of that province is approached with trepidation as the flow of dialogue begins to lose its fluidity and its effectiveness. In the end, the effort to produce some topical ideas to mitigate the fears is worth the challenge.

It may be comforting to still see the humor in a well-told joke, but that is small change if it takes forever to recall a name, place, vital word, or, heaven forbid, an important piece of logic from the depths of musty memory cells. If the voids occur in critical situations, the embarassment can be devastating. To avoid too many red-facing events, the forgetfulness factor should be addressed to save one's reputation and keep sanity intact in the aging cage. Repetitive memory exercises are a must in the cage. Taking on new lessons in new disciplines that stimulate the mind are one main way to keep the cells in tune. One creative challenge a year such as debating, writing, memorizing words and phrases from historical events, or learning or relearning to use a musical instrument will serve to keep the brain cells fresh. One very simple exercise is to pick a letter from a thesaurus and, while out walking, think of all the words in one's vocabulary under that letter, an exercise for both mind and body. If your vocabulary is too sketchy, check the thesaurus, make a list and carry it with you on your walks. Do memory work on the road.

Getting up in the mornings can be an experience in discovery. What new mysteries emerged overnight? For females, they have a different menu to consider. It entails basic stiffness of the limbs and back, night sweat residues, and the ennui of insomnia. With males, the list includes frequent urinary wake-up calls, nightmares, snoring, and general discomfort from the pressures that a heavy body can exert on legs and arms at rest. To view a ghoulish reflection in the shaving mirror every morning is not that invigorating.

However, the positive aspects of living in the aging cage can balance some negative aspects. If the prospect of a rain of falling hair on old shoulders is not overly worrisome, the silver of a manly mane can incite praise from some and envy by others. In the time spent outside the cage, one can relish freedom from small irritations. Not asked to dance at parties, opted out of being tabbed the designated driver ad infinitum, and evading attention of a personal nature by younger femme fatales, all add value to the credit side. The opportunity to share some time with compatible companions no longer on the clock can lift the spirits for many a day. The opportunities are many to be enjoyed within the physical and mental capabilities that prevail. The pressures of the negative side of aging and its fearful vision of a locked-in life will be diffused by a positive attitude and a strong will to stay with the changing times.

Super seniors will have realized early on that the fates seem more favorable to those who work to keep a positive attitude. Whether one has difficulty maintaining viewpoints that are devoid of rancor and malevolence, an overall perspective that sees light instead of darkness in the corridors of the mind will serve to be a key to open and close the aging cage at will. It is well to remember that we all have been given only one chance to live our lives on planet Earth. To waste that precious time in bitter recrimination and ill spirits will be called a pity at the golden gate. So it may be wise to see the aging cage as one last opportunity to right all wrongs, atone all sins, and view the future as a gleaming light of joyous opportunity.

Attitude is, after all, within the controlled purview of each human individual. To embrace any tendency to see and feel only the negative side of reason and events would lead oneself into dreadful conflicts of behavior in a welded aging cage.

Life should be lived with optimism and a comfortable state of mind, right to the end of the road. Every beneficial day of joy can add another to the potential aging stretch of time.

Readers Personal Space for Thoughts and Ideas

The Beauty of Words

Words that are assembled in poetic verse direct the messages into emotional regions of the human brain. They convey an aura of understanding that goes well beyond the commonplace. Poetic dialogue demands some mental effort to be appreciated in full and is therefore a tool for all cagers to consider in their dotage. To read and write poetry, even if it might sound doggerel to true poets, is not demeaning in any sense. If its utilization is as a minion of the mind, it may have power to extend useful hours in the aging cage, adding untold time to mental fitness and good years in a lifetime. Then one cannot logically question its poetic utility.

In a global sense, with so many languages in use, words can both converge emotions and expand understanding. Something is always lost in literal translation. The concept of a universal tongue is not practical in a multicultural world. The Tower of Babel is one story that illustrates the difficulty with a single vocabulary among a host of dominant personalities. The resultant chaos and outright conflicts defeated the strategy of mutual cooperation, and in despair, the ruling forces broke the unity into separate vocabularies. Current advocates of anti-multiculturalism use this Babel experiment to validate its value in lesser communities. However, lost in the emotions of the concept are the difficulties that arise from having to design and finance equal opportunities for all cultural groups to assure equality.

For aging cagers, the beauty of words can be a boon in the strategy of extending mental acuity. Words are the spice of life. They can be made to represent so many thoughts, actions, emotions, and reactions in our daily contact with the environment. They can color visual experiences, define exactly critical documents or terms of incarceration, tenderize or terminate loving relationships, spin truth into dissimulation, excite or diffuse tense events, create word pictures for readers, and generally influence most, if not all, human communications. If cagers are to benefit from word influences, they can be called to study the vocabulary of their milieu in greater detail and lodge as much as possible in the memory banks.

Solving word puzzles creates new mind neurons, which is beneficial. Creating new word puzzles is more dynamic, even stimulating.

The important virtue in words as neuron stimulators and memory refreshers lies in the necessity of rote repetition. Memorizing poems, names and addresses, capitols of nations, mathematical formulae, tables of telephone numbers, words of songs, dates of one kind and another, and passwords of so many things that have become commonplace in usage, are all techniques that challenge the mind to refresh itself as frequently as necessary. The process behind valuable memorization is not definable in common terms, but its effectiveness will prove the rule.

The easy flow of words in well written prose is a fact. Badly written text can be irritating and be avoided as a consequence. Good writers prevail. Bad writers fail. All readers of both prose or poetry become the judges of the magic music in the words.

Readers Personal Space for Thoughts and Ideas

The Brain:
Mighty Muscle of All Humanity

The human mind is the most powerful attribute of the human race. In a brain that is no more than three to four pounds of spongy tissue, the miracle of life is developed, its genre is ensured, and its evolution is promoted. A process of electromagnetic gravitational activity in billions of cells and neurons taking place every second for each lifetime is taken for granted as if it were mere bagatelle. It is the most intricate and functional mechanism in all the world. Humans should be amazed at its complexity, and awed by its precision and competence. The ultimate treasure given to all humans on planet Earth. It should be protected and treasured. There is no greater power that is contained in a space so small and has an influence so monumental. Think about it.

Concern with the state of the mind is a critical one with aging cagers. The onset of incapacity, if too early in life or too rapid, can be a most devastating consequence in the final phases of existence. With those who have consistently ignored the effects of bodily abuse in early years, the onset of senility gives no quarter to those that try but they rue too late.

The stages of brain deterioration have been studied, evaluated, and described over the centuries, but a clear picture of the way the mind processes incoming and outgoing data still obsesses, intrigues, and frustrates those engaged with trying to duplicate brain functions outside the human head with man-made technology.

The human brain is partitioned into mutually connected compartments. Each has its specific role, either singularly or in cooperation with each other in microseconds.

Two principle types of nerve cells occasion the transfer of data in the brain. One is called the neuroglia, and its function is to literally babysit the other type of cells called neurons. If a message concerning a situation needing attention is received from a sensory neuron in some area of the body, the nervous system passes it on to a process-neuron in the brain. The traffic of data moves from one such neuron to another through electrochemical fibers called axons that are the so-called traffic system of the nervous network in the body. A veritable web of dendrites and one electrically sensitive axon fiber surround the core of each neuron. The end of an axon may have a few or many fibrous offshoots, all with terminals that may have bulbs at their ends. The bulbs are electrically reactive and have a chemical coat. When the electrical message from the bulb is strong enough, the coat sends chemical messengers called neurotransmitters into and across the synaptic gap to link up with a correct dendrite in another neuron. This process continues repetitively in a chain of neurons, if and until the message reaches the central system and an action-neuron is activated. Not unlike the zero/one switches on a computer chip. The push the traffic enjoys as it is moved across and along the network at the speed of light mitigates the distance between neurons.

Aging cagers need not be overly concerned with the structures in the brain, but they should be aware of how their health/behavior can slow the degeneration of the neuronic system. Because the neurons in the brain do not regenerate easily during a human lifetime, the inventory is estimated at some 100 billion in each human. They wear out in time. There are two basic systems of communication in the human brain. The system that maintains the structure and viability of all neurons is called the neuroglia, or glia. This set of glial cells is important to the proper operation of the neurons and continues to divide during the entire lifetime of the host. The estimated number of neuroglia in the average brain is thought to be some 50 trillion, an incomprehensible number, and as such, at best, an unrealistic guess.

To keep the two major elements of brain function in top working form prior to and during aging, the cagers are well advised to understand that the process will rise or fall on the amount of exercise and good living habits that are used and maintained. Challenging the brain and body with exercise and problem solving causes the neurons to strengthen and will be directed to increase their capacity to fill the gaps left when some neurons have died away or been rendered ineffective through bad decisions by their human masters. Exercising the body has positive effects on the health of neurons. A healthy body begets or sustains a healthy mind. The types of physical exercise are limited by aging, so any plan to ensure conditioning must be selectively based on the physical condition of the cager. Walking is a primary exercise that should be part of every human activity, young and old. It can be scaled up or down in terms of effectiveness but grants many opportunities to keep it interesting. Exercise for agers should be within energy limitations.

Mental exercise can embrace many different venues. Some are active such as games that involve wide word usage and creation, new learning experiences with special focus, social debates with erudite competitors, part-time challenges in sports or teaching, and generally getting involved with some hobbies that demand eye-hand coordination. The principal idea is to challenge the brain to produce new slants and strain itself without worry. Once the decision to embark on mental exercise projects has been seriously taken, the neurons will accept the challenges of staying up to date, crisp, and productive.

Challenging, deep-thinking exercises create new neurons, a process once thought to be impossible. The hippocampus is the center where the brain enjoys the hard work of creating new neurons. In this set of writings, it must be appreciated that the purpose was not to pose an academically intense set of teaching lessons. It is an exploration of possibilities and thereby create an arena of discussion and a future dialog that may bring new insights into view. If it seems difficult to understand, the answer lies in asking oneself "what is the author trying to convey". Answering that question will arouse an interest to think it through. That is the main purpose; creating the need to think and that will exercise the brain.

Readers Personal Space for Thoughts and Ideas

Plumbing the Well of the Soul

As the realities of the aging cage keep plucking away at the lazier neurons of the mind, concerns with some of the esoteric concepts of existence become more clearly urgent. One of the more Delphic views that emerges for super seniors to engage their time and energy is seeking the place in time and space of the human soul.

Like all subjects that are floating in the space of mental awareness, this one embodies a host of controversial opinions in public debate. With no hard handle to grasp as a proven definition, everyone has a right, it seems, to grapple with the myths that fill the galaxy. Living with real prospects of their own fate, aging cagers may dwell deeply on the concepts of the soul to find comfort, some security, and a ready blueprint of the future as they face the fates that may be waiting out there in the afterlife.

Without a doubt, the human soul might occupy some space in every human mind. It is thought to be like a well that has an infinite depth. In very recent times, a concept for using cosmic space to store digital data has appeared and is labelled as

THE CLOUD. It opens valid conjecture that perhaps data other than that which is lodged there by humankind already exists from other sources. Could it be that the knowledge of all past humans is resident there?

Memory is both in the human brain and may have a great deal of knowledge stowed in cosmic space. The soul, is for all intents and purposes, the history of each individual that reaches back into outer space. So, if we are to be logical and defy the preaching's of the acolytes and minions of religious academia, we can begin to philosophize another blueprint for the soul. It takes no great stretch of the imagination to believe that the soul, ever human, living and dead, is retained evermore for purposes of its historical record.

Those living in an era with their aging cage have the privilege of tapping into their electromagnetic memories in the cosmos through subconscious contact in dreams and with the tactics of mysterious mental events. One may need to be a visionary human in order to perceive three dimensions of information and to defend their concepts of the presence of subconscious forces in their philosophical universe. Remembering that energy once created is eternal.

Because it is a known fact that human energy cannot be lost, this suggests that, when one opts out of the aging cage, locks its door, and waves farewell to one's mortal history, his or her energy transfers at the speed of light into cosmic space as electromagnetic particles, never to be lost in an evolutionary eternity. Though scientific fact cannot prove it, perhaps those cosmic particles seek and unite with their ancestral predecessors in the universe and become available to succeeding generations to color their time and creativity here on Earth. It could be called the source of creative, genetic, human effectiveness. Genetic influences must come from a progenitor, and that data may well exist in space for all time in an inventory of eternal lore.

While the concept of cosmic involvement may create questions and be an unproven theory, it is no less credible than the long established figments of imagination that have supported the theories of the many religious and political creeds of centuries past. The strength in every belief will bypass reality with great abandon, so an infinite calendar of hoped-for truths become the standard myth of heaven. With evolution on standby, however, no human concepts or creations can have an infinite life-term except in eternal space.

There is a logical reason in the Universe for the curtain of secrecy that hides the next world from all humans while on planet Earth. It keeps the future intact and prevents the many experiments of mankind from messing up the cosmos in the same way that their greed and search for fame have done on Earth. The human animal has not yet proved their validity as a worthy keeper of such cosmic facts. Perhaps one day, as human maturity evolves and all humans lose their propensity for selfish pursuits, the aging cage will see an open door evolve that limns another world, another life experience in those mists of cosmic haze that cloud our present view.

Until that day arrives, humankind must be satisfied with unproven concepts of the human soul. All theories must be taken at face value, and their authenticity will be exposed to the common sense of each and every human being. Tied as these ideas of the soul are to our human spirit, this eternal saving grace should exist in one's attitude toward others and, inevitably, to himself or herself. With the worldwide separation of cultural togetherness, tests of tolerance have failed. Learning to be a brother's keeper in matters that embrace the deeper values that reside in the subconscious mind are not yet accepted principles. If they never will be, the hope for the opportunity to view heaven on earth may not be attained before any usefulness of cosmic development vanishes, along with the genre of the human creature from this universe.

Many of the genres of life that came to be when the Earth was first created from the cosmic stew of the universe have, over millions of years, evolved themselves into oblivion. They failed their ordained missions. No longer coping with the pressures of change, they were committed to extinction. Perhaps their souls reside in cosmic space.

Readers Personal Space for Thoughts and Ideas

The Human Mind: Cosmic Communicator

It would seem logical to marry the brain and mind in the same basket as they do operate in full conjunction with each other. However, in the interests of clarification, it also seems logical to separate them as two elements in one body.

The brain is the mastermind. It contains the central nervous system that is the computer of all applications in the body. It performs these controls and instructions through a system of nerve cells that are wired electrically and/or chemically to sensor cells in multiple sites in and on the human physical frame. The communication nerves are known as neurons. They are of three types, each with a specific task. They are:

A. **Communicative:** The neurons route the messages from every part of the human body that signal some action is required either to assist, repair, or remedy a perceived or subliminal condition.

B. **Activator Motor Neurons:** They are connected to all muscle cells that make things happen. These cells have reactive capabilities in microseconds as they direct and track every change in muscle needs and positions.

C. **Neuroglia:** They babysit the functions and health of the other neurons. They also control the movement of data across the synaptic gaps that keep the neurons physically separated by changing electric signals to chemical when some event is calling for action.

The brain controls the human body with six main types of basic electric and chemical tools: neurons, motor muscles, epithelial sensors, blood distribution, skeletal bones, and tissue connectives. They all are systems that the conscious mind experiences and provides the necessary motives and actions once a conscious decision to act has been recognized and reached. The decisions are most often formed with the full knowledge of the decider using personal judgment with an objective event or idea.

The secondary system in the brain is the subconscious one where the needs of the human body and mind are fulfilled outside of normal cognizance. All the activities of the body that are performed automatically twenty-four hours a day are the purview of the mind as pursuits that must occur to ensure the continuation of life itself. They are etched into the system but act instinctively. Just thinking of their intricacy is mind-boggling.

The function of the mind is the eternal mystery of humankind. We know what it is supposed to be, but how is it created, and where from does it originate? This quandary has bothered some of the learned world experts in the field of brain intelligence for centuries without any definitive philosophical answers.

Subconscious triggers, entirely in the brain bring new ideas and imaginative events to be recognized and modified by the mind. Parapsychologists challenge this theory today. They believe that such reactions cannot be self-engendered and claim that definable motivators can be tabled as follows;

- **Emotions:** Love, hatred, fear, envy, rage, greed, jealousy, joy, affection, and ennui
- **Beliefs:** Philosophical, religious, and personal
- **Introspection:** Study, prayer, and meditation
- **Character:** Integrity, compassion, generosity, cupidity, dissimulation, intellect, and greed
- **Pressure:** Health, ego, competition, and boredom
- **Security:** Safety, awareness, and retirement
- **Survival:** Mortality, risk, opportunity, and challenge

The reasoning that the mind can independently bring new ideas into focus from thin air fails the test. Nil input means nil output is the mantra of the disbelievers.

What has been proposed, but has not been generally thought credible, is the theory that the cloak of electromagnetic cosmic particles that surrounds the Earth just might contain packages and streams of creative energy that impact the human mind and give it the impetus to kick-start something new and unique. Time may mature the human mind enough someday to understand the influence that the aura of cosmic energy that fills all corners of the Universe has on all the living creatures of the Earth. As they are constructed of electromagnetic energy themselves, humans are all coated in the invisible cosmic force field without knowing how and when one might experience its influence and impact.

That question occurs when humans use the power of prayer to seek divine wisdom and assistance in their search for some absolution and meaning in their life. The use of godheads to identify spiritual energy is merely using identifiers that the human mind can relate to as a known type of force within their levels of conscious cognizance. The instinct of recognizing that a power greater than that which is obvious, felt by most humans, would be, and is in some quarters, the first step in accepting the presence of such forces in the Universe.

To make it possible to be managed as a bundle of force particles, beliefs that can be seen with human cognizance are yet to be discovered. It may well be that the subconscious forms of dreams and unprovoked instances of magic will one day mature into force units that can be utilized, if not controlled, by the efforts of humankind. Its cosmic intentions may become another virtue in the lives of humankind. It is an exciting concept for the future but yet too farfetched in a

Universe that does not recognize the clockwork of human time for many generations yet to be born.

It can be logically conceived that, until the human mind can accept without doubting that the cosmos wields a constant, albeit subtle, influence on the subconscious patterns of human thought in mysterious yet recognizable ways, direct contact with the energy field of the universe will not occur. Just how many centuries must pass for that level of spatial thinking to become a recognized and available resource is beyond logical prognostication. However, if the forces of evolution are to be believed, it will happen sometime in the approaching future, but perhaps not in current Earth time frames or even cosmic multi-millennia. To miss that personal power is a regret for all thinking humans to bear for their lifetimes here on planet Earth.

Readers Personal Space for Thoughts and Ideas

Memory Loss: Damocles Sword (Omen of Disaster)

Of all the debilitating ills that come along with the fading capacities of the neurons in an aging brain is that of the loss of memory flexibility and capabilities. As inevitable it seems as death, the inability to remember with a clear and lucid mind is frightening and immobilizes a flexible mentality. Not only does it signal the potential approach of Alzheimer's and Parkinson's diseases, but its day-to-day importance in our lives is seriously compromised. With the tool of communications denigrated and the power of speech reduced, a new set of coping strategies is ordered for those in the aging cage. By the time the memory loss is getting severe, it is too late to try to use ameliorating tactics to stop or slow down its creeping effects substantially.

Memory is a positive force, but its composition is vague and mysterious. It is known that there are two kinds of memory. The first is long term, and that pool of data is filled with information and experiences that make up the learned inventory of one's life.

While it seems somewhat incongruous, this side of memory is that which lingers longest in the useable recall cells of the brain. It must be concluded that the primary reason for this anomaly is centered in self-confidence, experiential drama, constant upgrading by the mind, and a good neuronic monitor. It decides what incoming data is important enough to warrant storage space for longer periods of time. These remaining long-term memories are of great importance to the cadre of aging cagers who are immobilized by physical disabilities and must turn more to their minds to find worthwhile meaning to and in their lives.

The second type of memory is called short-term recall. That part is the up-front working tool. It is in constant use throughout normal conversation and where short-term issues are in question. Names, places, current facts, remedies, tactics, feelings, attitudes, skills, passwords, and many other requirements for memory are usually brought up without any more reason than a possible need. Because short-term memory demands are usually action-based, once they have been utilized, they are discarded or stored, if vital, in the long-term memory cells. Perhaps because of this factor, short-term memory capabilities are forgotten in short periods of time. Without some restorative exercise, they will gradually lose their energy and are discarded without traces left in place anywhere in the brain.

Scientists using brain scanners have determined that, if the use of brainpower does not die with advancing age, the residual neurons increase their capacity against the grain of normal deterioration. This limns the fact that, to slow the aging of brain and memory cells, one should keep the need to supply information from the conscious to the subconscious mind as actively and more energetically than ever. The brain reacts to the needs, and this stimulates the memory cells to stay sharp.

Educators in public school systems once taught memory work as a matter of course. The alphabet, days of the week, months of the year, and times tables were taught as rote exercises. Most of those, once remembered, are never forgotten. But constant recall will continue to be required so that aging neurons will stay strong and active. Once etched into the brain, the long-term memory cells become available on a moment's notice and will act like short-term memory when needed. The new age has made the rote learning system obsolete as electronic computing has overtaken the need for human memory files for addition, multiplication, division, and all forms of mathematics, structured thinking patterns, and new inventions. With computers, they no longer drive the mind to recall formulae, and the challenge to the brain is minimized to the point of making sloths of neuronal cells. The ebbing rate of unchallenged memory muscles accelerates aging. Science may be a vital component in human life, but it has served to reduce the memory capacity of human brains to a serious degree and created thinkers into audiences.

To ensure the productive competence of memory cells, the time spent in the aging cage doing recall exercises can be profitably used to extend the years of retirement pleasure. Some of those exercises have been mentioned in earlier notes and writings, but the potential for new methods is extensive. It only takes the impetus to start the thinking processes and experiential exercises to turn ideas into worthy stimuli. To use the power of computing, digital apps with mentally challenging exercises are available, are easy to use, and can exercise the brain.

The mind exercises must have a healthy quotient of difficulty. If the exercises are mere novelty, the impact on neural retention, improvement, or replacement will not be effective. The mind must attempt to reach new ground and undertake hard work to conquer learning tests of any kind. These might be in puzzles, music, art, writing, poetry, exercise, and any other trials that will work the short-term memory with taxing trial. Not to totally exhaust the normal energy inventory, but as nearly as possible. The challenge for those in the aging cage can be inspiring.

Readers Personal Space for Thoughts and Ideas

The Genetic Trap: Hard Rocks on Destiny's Road of Life

It can be surmised that much in the origins of each human destiny can be traced from one, two, or both major inputs:

- A personal gene pool reached through parental coupling and back through them to a dim, historical family tree.
- The genetic elements that devolve from long-term historic genes, plus those that are infused into the pre-borns and infants from the memory resources of the electromagnetic cosmos that surrounds the Earth. Exposure to the cosmic environment is real, and newborns act as blotters from this source if they are healthy offspring.

The genes that are carried in the sperm and ovum of the hosts are known factors that can be traced from parents to their offspring. Those that may be added from the Universe, and this possibility, moot as yet, could supply an alien gene that would be dominant enough to create new personalities and talents. There must be some logical explanation for the emergence of raw talent in so many fields of existence here on

Earth from parental genes that have no evidence of such latent tendencies. The fact that such events occur might establish that electronic particles or threads from deceased humans do return from storage in the Universe. That they may intervene during the pregnancy and birthing processes of human mothers here on Earth may be under scientific study in labs of erudite explorers and scientists around the world.

A significant part of the genetic trap lies in the reality that parental habits prior to and during pregnancy can affect the characteristic nature of their offspring. Two noted indicators are the pass-through genes that have been, in some way, influenced by tobacco and alcohol usage by either or both natural parents. And the effect that some sexually transmitted diseases (STDs) can have on male sperm efficacy, in the long term, have not been traced, but logic suggests it may be possible that they infect the offsprings. The consequences of parental indiscretion in not protecting their future human progeny are often never noticed until the damage is recognized in the infant or adolescent. By then, it may be too late to gain some leeway in the fight against lethal natural forces.

The genetic trap is created and sustained, perhaps even reinforced by the behavioral and communicative habits of the parents at some stage in their developing maturities.

The first of these is the lack of understanding of what their pre-union, uninhibited lifestyles might manifest in their offspring. The trend to free expression in all facets of life and those who find lucrative returns from fostering such behavior are those who may arm the triggers of the genetic trap. With both parents, those putting forth the seeds of super satisfaction from a lifestyle of dissolute freedom, the destinies of the offspring have no personal defense.

The second invidious component in the genetic trap lies in the lack of rules to guide young parents in their roles of parenting. The event of childbirth is traumatic in every sense for those who are serious and not frivolous about the sanctity of child creation. Young mothers who may have trailed or opposed their own materfamilias will either copy mostly all the erroneous practices or will discard the experiences outright and cut a new, but inexperienced, path to parenting.

Without guidelines, mothers who have been exposed in traditional marriages or via unplanned pregnancy with or without a responsible male involved, may tend to be too permissive with their offspring. They can destroy the development of respect, responsibility, and compassion in the child. Caught without any but loose ties to the reasons for their existence, the genetic traps snap shut on unprotected offspring.

In order to escape the clutches of the trap, the formation of a personal set of worthy qualities seems very necessary. Without the stability of a parental home to pick them up when they bounce off bad mistakes made in growing up, many offspring may lose themselves in dissolute lives, illegal acts, promiscuity, and mostly unproductive existences until they choose to end it all or merely sink into despair and insanity. With no lifesavers, they are lost. A loss of talent without just reason. With evolution an on-going reality, such losses, however minor, will, in future time, diminish the creative talent of planet Earth with perilous long-term consequences. **Creating new lives should not be an irresponsible nor a frivolous pastime. It places an incredible onus on the female who must decide on the virtue of denying her inborn urges to procreate against that of weighing the cost to herself and her community when the decision to create and mother her offspring is at hand. New lives introduced to a dysfunctional environment for their maturing lives is one of the major social disgraces of the modern era.**

Readers Personal Space for Thoughts and Ideas

Energy: The Sneaky Thief of Waning Time

All those who are occupied keeping the aging cages fully shipshape realize, often sooner than later, that something is happening with their energy inventory. Without any warning, the quantity and quality of the go-go juice just is not up to the demands as it once used to be.

The ultimate paradox is that it deteriorates at a constantly escalating rate in an unheralded process that will catch the unwary cager with his awareness down around his knees. Getting out of bed in the morning dawn lacks bounce and maybe even throws disorientation into the mix. Suddenly, the world around seems to be reeling, and balance is in jeopardy.

Muscular actions that were once automatic must now be closely monitored to ensure their proper utilization. Instead of using muscles with subconscious traction, they tend to sag and must be specifically directed in most of ordinary tasks. Walking paths seem to take longer to traverse, and their selection demands shorter distances to conserve that fading energy. Interest in the surrounding beauty of the environment is leached away by the need to watch where one is stepping to avoid bone-breaking tumbles.

Sleep time is reduced and, for males specifically, is broken up throughout the night by necessary visits to the urinal. Thereafter to cajole Morpheus into renewing his responsibility becomes a battle between a mind intent on reviewing dialogue and a body that might have a desperate need for rest in a rumpled set of sheets.

Entering a room with a clear purpose in mind and having forgotten the purpose of the trip on arrival is a common lapse in memory experience.

Understandably, aging cagers must adapt to their waning energy supply. The transition from an active lifestyle to increased passivity does not sit easy with the senior segment of society. With higher levels of wealth, money can buy many other activities to pamper lower energy assets. However, those passive pursuits may spell shorter lifetimes if they fail to challenge the mind and fail to exercise the body. Entropy is just around the. corner and does not hesitate to co-opt the destiny of all those flaccid humans.

Energy, as understood by scientists, takes the topic well beyond its importance to the aging cager. In these masterminds, energy is seen as the keystone of life as we know it. In the writings of thousands who have tackled the problem of defining this unseen but vital force, there are but two major categories of energy. One called kinetic has many subordinate types and is the power of motion. Energy is so important to human life that Homo sapiens could not exist without its influence. Any body of mass that is moving is said to be doing so under the force of some variety of kinetic energy. Be it mechanical, electrical, environmental, chemical, thermal, human, or any force that moves mass does so with the virtue of one form or other of kinetic energy.

The second category is called latent or potential energy. This defines the fact that cells in the human body, batteries, microscopic electronic chips, equipment, and all other normally inanimate objects can store energy and hold it for differing periods of time until it is needed. Power sources use and create energy, and they are many in common applications.

A major third category is being seriously considered but not yet well recognized as an energy source. This one is called universal or cosmic energy. The cosmic environment of the Universe is filled with electromagnetic waves and particles. These waves, mainly from the sun, are positive energy and permeate the cosmos, the planets such as Earth, and all its other living and inanimate inhabitants alike. Humans, like all flora and fauna, are bodies of electromagnetic energy that is both kinetic and latent for every second of our living existence. The process of creating energy is eternal and operating within the billions of cell contents as atoms collide, evade, and attract each other every living second of any lifetime. The elixir of life is energy. To keep it active and adequate to the needs of aging cagers is a passport to an existence that progresses fully to the very end of one's blueprint of destiny.

For aging cagers, the creation, retention, and wise utilization of energy can become an obsessive challenge. While it is appreciated that exercise of mind and body are both vital to the viability of the energy inventory of the human creature, it calls for some disciplinary behavior to fulfill. Several natural forces oppose a regime of energy management:

A. **Antipathy:** This sense of hostility is a form of self-reproach among aging seniors in the throes of retirement. It is not easy to avoid cursing the losses in normal mobility and precise flexibility that may have been taken for granted in the maturing years. If a good regime of active exercise is negated by a war within by the two antagonists, motivation and antipathy, the years will pass rapidly and fruitlessly too soon.

B. **Apathy:** If the mental battle between the attitudes of "can't be bothered" and "get up and get with it" is won by the wrong set of forces, the case for exercise in or out of the aging cage will not prevail. The insidious invasion by the troops of apathy are present everywhere and must be defeated by the force of human willpower if the blueprints of good health in the retiring years are to be kept current and active.

C. **Physical Disability and Fatigue:** The loss of vital parts of the body through disability can extend the difficulty of maintaining an active anti-aging program. However, with modern science gaining on these negatives, the possibility of a dynamic program can be maintained. However, overexercise can become the primary cause of fatigue and must be monitored effectively so that excessive use of residual energy does not impair mobility.

D. **Training and Financial Resources:** Private and governmental services to advise and aid in setting up and maintaining good exercise projects are available for those who are seriously interested in fighting the foes of aging ennui.

Energy use and storage in the human body is dependent on the wise and adequate consumption of food and water. A good diet, planned by experts in the field, can be an important asset for an aging cager who may need optimum regimes of eating and drinking to retain a workable inventory of personal energy.

Ode to Joy: Tuned for Super Seniors

Beneath a fiery sun above, we shape our own obsessions,
> To seek eternal peace and love and make each day's
> concessions.
Musing not nor why nor where, the fates make their decision,
> To keep all humans here and there, well, in a state of
> fission.

That elusive pixie of joy, that tantalizing sensation of well-being that engages us too rarely, is forever a challenge. Our dream is, of course, to don that garment of entrancing emotions as fully and as often as possible to color many a mundane day's experiences.

All humankind seeks lasting joy. It is not a selective privilege of rich or poor, famous or infamous, black or white, or young or old, so each sector of society must seek its hypnotic embrace in its own singular way. That search is of particular interest to aging cagers who may have more hurdles to surmount than do their younger aspirants.

So, from a cager's perspective, what might be some key tactics to ensure that an older mind and body may have a fighting chance to reach satisfactory levels of joy against the negative perspectives of a critical and judgmental coterie of less mature contemporaries? There are some major shifts in awareness that might be practiced as one makes the transition from an active to a passive phase of living. Some of the more important are seen to be as follows:

A. **Respect for Others' Privacy:** This rule of behavior does not ring a resounding bell in the minds of many retirees. If one has been engaged in environments where interfering with co-workers' thoughts and ideas has been a rule and often encouraged, doing so in a social context is not appreciated. It is good protocol to respect others who have long-established routines and not engage in controversy over whether they should change to adapt to the provoker's raison d'être, or systems of existence.

B. **Compromise:** The art of compromise is never taught. It must be learned in one's own mind. Accepting the validity of every other human's right to live according to his or her own personal rules is a logical start. Accepting other viewpoints is fair game for argument, but criticizing the authors for whom they are, is not. Without compromise, there is little joy. Meetings of the minds and spirit with others is one essence of true compromise.

C. **Attitude:** When entering the confines of the aging cage, a healthy attitude about the promise and potentials of the future is a mandatory condition if the joy in that experience is to prevail. With a serious disability, this can be difficult. However, there is no real virtue in imbibing of the vial of misery as a state of mind when it is, in most cases, somewhat controllable by its possessor. Humans are conditioned to react positively to a fellow human with a happy face regardless of the condition of the body. Dealing with the negatives of aging in a positive and joyous manner will make the precious days of life on earth an eternal joy.

D. **Companionship:** Sharing quintessential time with like-minded humans is one road to real joy in the aging cage. It sometimes takes a lifetime to discover just one such companion. The panorama of acquaintances can be counted in the thousands-plus, but most are passing fancies that rise with the dawn of any new day and wither ere t'is over. There are many reasons why so many fail to gel into true companionship. With men and women, the wraith of envious competition is always there. The insecurity that voids the opportunity never lets the freedom of emotions to develop. With a real friend and companion, absolute honesty does temper the emotional breezes that swirl across the plains of trust and confidence. Secrets should be respected and kept sacrosanct. What one learns and knows must not be used to gain advantage over a companion for ego or other reasons. Equality of status is to be based on human virtues, never on possessive attributes. Affection of an honest kind is the touchstone of a joyous friendship. And there can be no hidden personal agendas on either side if a joyful companionship is to grow and sustain over time.

What Is Life?

What is life? We may ask in a song.

Here for today. Tomorrow it's gone.

No chance to relive it. We would if we could.

Do you think we might change it?

Let's say, "If we should?"

The days putter on and grow into years

With joy and elation and even some tears.

For the young, they are endless to buy, beg, or borrow,

As if in such minds there is no tomorrow.

When shadows and mist start to dim what we see,

It is time to discover reality

Then examine the past. That's what memory's for,

While fate now envisions the last golden door.

But whatever is life, it is surely worth living,

To care and to share and focus our giving.

For aging cagers, after super seniordom is visible on the horizon, time is no longer measured in years. It shrinks to weeks and days and may even be hours. So the meaning of life is to enjoy, yes, enjoy, every moment. It is supremely precious and irretrievable. Use it well and as fully as possible.

Readers Personal Space for Thoughts and Ideas

Super Seniors: Pros and Cons of the Golden Age

The final missive in a series is generally the hardest one to write. In the Exx-Rays group, so much of what is still possible to examine will have to be left behind. But, with all examinations of a pure and speculative nature, particularly those that are written not to be instructive tracts but to be stimulators of unconventional thinking processes, a stopping point must be established. With the hope that from the ideas thus planted, the mental soil has been diligently tilled, and the promise of a crop or crops will show up in the produce of many maturing minds at some time in the earthly future.

So what might be the message from the Exx-Rays material? The age of super seniordom is unique and special and can be a most exhilarating phase on the voyage along the roadway of one's destiny. The joy is not akin to that which one experienced while dancing with the hubris of one's youth, but perhaps is a more meaningful and serious enjoyment that may be muted just a bit now and then by subliminal awareness of the unknown challenge that lies ahead.

Super seniors, in the main, have reached the point where usual concerns and many driving motivations are no longer seriously relevant. The search for adequate security has generally been successfully forecast and fulfilled with savings, investment plans, and pension options keeping the wallets filled well enough. Simpler cash flow management or use of money managers will ensure financial viability until the need for them no longer exists.

And equally important for a sublime, golden years existence, the ego-driven goals of fame and fortune are no longer weighing quite so heavily on one's lifestyle perspectives. That, in itself, makes for less envious and emotional reactions when viewing other socially apparent human subgroups. This, in turn, creates a warmer climate for new friendships to develop and previously only dreamt-of experiences to be enabled. Not quite heaven on earth, but possibly as close as anyone can hope to get or attempt to predict.

The reduction of intrusive cogitation from sexual angst at high boil to subtle simmer may be both a bane and boon. Most of the super senior coterie have set aside attempts to re-create the emotional ups and downs of youthful dalliances by using trophy brides and sex trade sirens and have accepted whatever levels of atrophy has occurred in one's manhood artillery. Borne with the usual equanimity and acceptance that lost causes eventuate. One rarely hears any sighs of regret from the distaff side who were early partners involved but are now stakeholders in the downslide to complete abstention. So life goes on until the call is heard and one faces the challenge of a new experience where one's beliefs may well be tested.

Eulogy

How soon bouquets of flowers die,

No longer kissed by dew.

Fate's arrows seek the errant heart

To pierce it through and through.

When memory fails, the spirit wails,

"I'm lost, caught in the past."

When life ebbs out, the soul may shout,

"I'm free, yes, free at last."

Amen.

WAS/2012

Readers Personal Space for Thoughts and Ideas

Cosmic Cybertreks (Prologue)

Cosmologists, physicists, mathematicians, meteorologists, and scientists of all types are said to have a reticence to view events and ideas that do not fit their preconceived notions and proof-laden attitudes and ideals. They are said also to be very anti-responsive to new applications to old concepts and thoughts that do not fit a scientific box and are thus generally ridiculed and rejected.

Despite this type of professional bias, thinking on different platforms has been emerging over historical time. Evolution will prevail, and this has been the force to sweep aside the negativism that could stultify creative ideation proposed from the minds of unusual and intelligent thinkers.

Approaches to the impact that the energy-laden environment around planet Earth has on human minds begs for some radical and innovative ideation. Much has been learned in recent years, but the consensus is that the human race has much yet to learn about the mysteries of the Universe and its connection to humankind. The maturation of the human brain is still in process, and as it reaches new and rarified stages of logical development, many new concepts of the gold still hidden in the cosmic mines will be prospected and exposed.

The cosmic cybertreks explore some of the interesting ideas that are literally foaming in and out of the mind of this one cosmic prospector. Without the confines of most scientific protocols, this mind feels free to wander into the virgin wilderness of unsubstantiated events and undiscovered atmospheres in the spatial void. The challenge is to approach the task with some logic as the guiding light on a pristine pathway with no clear road signs as yet established. The axes of criticism may appear, but the traditions of healthy controversy have brought fresh ideas into the spotlight for many years by those who fear not the sharp blades of censure.

Let the rains of criticism fall where they may. If it irrigates other creative minds, so be it. Sometimes the crop is a worthy erudition.

Readers Personal Space for Thoughts and Ideas

Cosmic Cybertreks Contents

Reading Sources for General Information

Stephen W. Hawking, *A Brief History of Time*

James Gleick, *Chaos: Making a New Science*

Benoit Mandelbrot, *Fractal Geometry*

Isaac Newton, *Principa, Book 3, Gravity*

Albert Einstein, $E=MC_{(squared)}$

Collins Knowledge Series, *Living World of Achievement*

H. Margenau, *The Scientist*

Michael S. Sweeney, *The Brain*

S. Goudsmit, *Time: Man vs. Clock*

Carl Sagan, *Planets: Solar System and Beyond*

A.C. Clarke, *Man and Space*

M. Wilson, *Energy: Prime Mover*

J. Pfeiffer, *The Cell: Bustling Metropolis*

Deepak Chopra, *Ageless Body, Timeless Mind*

Evolution: Eyeballing Space Effects

To address the subject of evolution, it is necessary to realize that two major forces are at play. The first is events that are occurring in outer space and over which Homo sapiens have no known influence. Evolutionary changes generally occur in eons of time and are beyond the knowing experiences of any one general lifetime. The second is the evolution that is directly influencing our planet Earth and over which we may have some influence and can measure the impacts that occur quite arbitrarily. Most can be experienced within the average life span of every human generation.

The subjects are fascinating and intense. They will be thought out separately in the interests of a wider group of spatial and obtuse thinkers. Of importance is the fact that every inch above the surface of planet Earth is not the firmament of Earth. It is all cosmic atmosphere.

Cosmic Evolution: Nature's Force for Change

This field of study is, by its nature, somewhat speculative and must rely on logical conjecture to support the premises that are proposed. New thinking in the world of cybernetics is structured methodology and will cross established myths and beliefs so they will invite challenge, criticism, and ingrained, often bitter denial.

Human views of the Universe are seen as a plaque of time and space with limits established for a start and finish of events that have occurred and are occurring. The start of the universe that we inhabit had a beginning said to be, in scientific circles, as the Big Bang. There is much controversy over this event as the starting point for the origin of the universe for one, currently unresolved reason. It has been said to have been a big bang start, which was a huge explosion of a "singularity." This word is much explained in scientific cosmological literature, but in short, the blast occurs when cosmic elements have reached a state where they overpower or are overpowered by the antibodies that control their growth rates. However, there are those who make the assumption that the Big Bang was a religious event ordained by God who planned and triggered the explosion.

Common logic defies this illusory concept. However, to satisfy the critics who opine that the big bang must have had a logical basis in order to have occurred at all, is extant. What is puzzling is that very little, if any, scientific study has been done to examine the concept of the Universe as other than a space in time. But because energy cannot be destroyed and is utilized, modified, or stored, it is a sterling factor in the evolutionary events that evolve in the cosmos, and alternative views must be recognized. Changes do and have occurred in current known time frames.

Because space atmosphere is infinite and time is eternal in the universe, they have no relevance when visualizing the dimensions of one and the longevity of the other in spatial thinking. This suggests that cosmic space should not be viewed as a panel of human thought, but be rather considered as a panorama that is constantly evolving forward, albeit over millions of earthly years. This view paints a picture that is a continuum of events that stretches back to an eternal space, through our visible universe in the present day, and will carry on into the eternal future when the time is ripe. This theory solves the question that creates doubt in proponent minds of big bang theorists, because it can be said that the elements that fused to cause this explosion were amassed in the earlier stages of universal evolution. That too perhaps was created by a previous big bang event. If big bangs are universe creators, they must have existed eons ago and may prevail for eons to come.

A primary phenomenon in our current universe is its constant, unexplained expansion. Could this be a presaging development that may be preparing another separate universe to follow that which we now occupy? What is there to prevent another big bang explosion that would end this universe and create the next one in the continuum? Black holes could swallow all the detritus of a dead universe and cast the energy back into the void. If the laws of evolution are to be believed, there must be a change when this current condition matures. With time being eternal in the universe, it may take multimillions of man-made years, but no one in the present day has any cause to care, obsessively, at its future inevitability. This concept is questioned later in this cybertrek series.

Planet Earth: Evolutionary Signposts

In the many writings on this subject by pro scientists, the focus seems to be largely on the function and effects of evolution as a normal event in the daily lives of all things on planet Earth. Evolution is seen as a consequential change in the status quo due to emerging events. The singular direction of evolutionary change is like an arrow that points in one direction only and is progressively forward from any steady state. The force that impels it onward is centered in the Universe as a natural law that cannot be denied by humankind and cannot be reversed. Humankind can, however, influence the stability of events in the forward flow of evolution on planet Earth. We do so by ignorant exploitation of the terrain that (a) covers its surface perimeter and (b) blankets its spatial atmosphere. Because of human tendencies to worship fame and fortune, this exploitation is extensive and pervasive in a brainwashed, competitive culture, forever seeking out those roads to glory.

The sun is the chief source of energy on the earth. However, only less than 5 percent of that availability is used on this planet, and the remainder is distributed in the cosmic atmosphere that embraces the universe and surrounds the earth. This energy also permeates all animate and inanimate bodies so all are as one with the energy in cosmic space.

Cosmic events that are changing planet Earth are irregular occurrences but have dynamic effects. Volcanoes have ruptured the surface over centuries and still continue to pollute the atmosphere with ash and gasses that have untold consequences over time. Earthquakes occur as warnings that the earth is shrinking under gravitational pressure, causing changes in the terrain in the spaces of land and water and perhaps in minor ways with the ultimate size of the orbit and its rotational speed around the sun. Evolutionary changes in these forces will affect, in time, the stability of temperature, climatic events, and living conditions worldwide. Tornadoes are another prime example of the power of climatic change, as is the tsunami that creates the change.

Rife and powerful sunspots bathe the earth with solar energy that, if unprotected, does cause short and long-term damage to the health of humankind. What damage is done to other genres of living creatures is immeasurable, but we place our trust in nature to adapt as needed. Historical evidence suggests that species not adaptable become extinct in time and are replaced.

Human interference in the earth's evolutionary changes is a classic example of deliberate or uninformed usage of natural resources. Overexploitation of nonrenewable resources creates deserts that change the balance of nature. This causes loss of freshwater availability, and because pure water is the elixir of human life, its shortage will restrict population growth and viability. The overproduction of greenhouse gases tends to destroy the protective shield called the ozone layer. This layer in the atmosphere protects, by reducing the penetration of excessive ultraviolet rays from the sun that are harmful to the health of humans. Atomic and hydrogen weapons have the power to kill all living creatures on the Earth. Their control is not guaranteed because criminal elements may use their availability to possess the power of fear over mass populations. Atomic power to create electric power has hazardous potential and demands a supreme quality of caretaker control. Instances of damage by the collapse of control have been and are being written.

It is believed that cosmic energy affects human minds and bodies, but because there is no acceptable proof, it raises questions about its existence. Because it coats the earth and permeates the human body, there is strong belief that there can be no escape from the influence of cosmic energy. Many humans tend to revere prayer as a method of appealing to a higher power for support and security. The deities of religious orders are seen as humanoid personifications of this power, and though logical proof cannot be claimed, ingrained established dogma subsumes any doubt. This begs a question. If humans find comfort and attributed rewards from assuming strong beliefs in such unproved theories, then can it be so difficult to think that cosmic energy can serve the same needs? Perhaps religiosity and cosmic ideology are identical forces on planet Earth.

Believers in the influence of cosmic energy on the human body and mind find more logic in the possibility because there is no barrier between the energy of the universe and that within the human frame. They are a unified, homogenous reality.

Miracles in the religious milieu are said to be "divine intervention" while those in cosmic quarters might well be called "pragmatic allusion". One definition need not compete with the other, nor denigrate its powerful effect. Human beliefs are sacrosanct and personal and therefore privileged to each human as a right.

A powerful urgency to act, arriving in the human mind when faced with a lethal event, is felt to be a cosmic signal to protect the existence of the human from a threatening death with, for example, an inspired word or idea. Timely at a moment of urgency, it is thought to be the influence of spatial knowledge projected into the human subconscious mind. These forces are the basic root of pop-up inspiration that motivates a genius in music, poetry, literature, science, and every creative example in other fields. Random speculation. to be sure, but worthy of some deeper thinking time.

William Salo

Readers Personal Space for Thoughts and Ideas

Pragmatic Allusion: Cosmic Communication through Cybernetic Symbiosis

In the workaday world, where most humans are domiciled, the use and understanding of upper-crusty words are rarely experienced and often resented. However, in those milieus where their use is daily bread and butter, they are very specific in what they attempt to communicate.

In scientific and philosophic arenas, cybernetics has several interpretations. In the whirlpool where ideas swirl, the intent is to succinctly define sensations and how they mingle in the human mind and body. The application of principles, together with methods of verbal communication, definitions of feelings as opposed to facts, is the intent of the word.

Symbiosis, in the professional worlds, is understood to be the defining word for the way in which two organic or inorganic entities relate and communicate with each other in Earth's environment. This does not exclude the ways in which units such as electronic systems, chemical components, social humanity, and flora and fauna of all types communicate with each other.

Experience and teachings condition humans to be mainly pragmatic and to believe that, if they can see, hear, smell, taste, and touch, these are the only components that define human visions of reality. While some figment of acceptance exists in that the electromagnetic fields in the cosmos can perhaps influence the human mind and body, it calls for too deep a philosophic intelligence to believe this force affects physical, everyday life. And yet, one never can be sure. Admitting that human brains are very complicated, very powerful processors beget examination of some extraordinary events that can occur along the roadways of life.

The brain has two thought mechanisms: the conscious mind where human awareness of all events is evident, and the subconscious mind that operates with all the automatic life forces, the memory cells, the neurons that control the input and output of information throughout the human body. Just what is and can be transferred back and forth between the human brain and the electromagnetic atmosphere of the universe is subject to some creative imagination. The basic need for survival, as a responsibility of the subconscious mind, is a given, but questions with no real answers emerge when attempting to track the sources of abnormal or impromptu events that occur peremptorily. Such experiences occur as flash ideas, evolving suddenly to artists, scientists, marketers, musicians, writers, poets, designers, and all who have stressed their minds deeply enough to solve new problems, create new methods, design new visions, and seek answers to emotional obsessions.

Many humans claim to have had such intrusions. In most cases, these pop-ups seem to appear as a consequence of some deeper cogitation that has focused the mind on the need and searched the memory for solutions without success. If the need has been seriously enough implanted in the depths of the subconscious mind, not as a command but as a plea, after an instant or short time frame, a solution or answer will pop up in conscious awareness. Some would believe that it is the result of rapid access to one's memory cells. If the memory has an adequate inventory of historical similarities or experiences, it could be the source of solutions to the needs. If not, the answer may not be so simple. They might be defined as an example of pragmatic allusion where the answers come to light out of the cosmic void.

Akin to divine intervention, one may seek the reasons for any pop-up event in a wider landscape of human topography. If energy can never be lost or discarded, then it must be reused or stored. Energy is power in motion. The sun's energy that is not utilized is dispersed in the atmosphere of the universe. Human energy, developed during a lifetime, is transferred back to the cosmos as particle bundles when a person dies. Those particle bundles are probably electromagnetic atoms that will carry quarks or mesons of information, imprints of knowledge, and experiences by all humans gained during their life spans. This data then becomes a part of the memory vaults of the universe used to meet a need appearing therein as a pragmatic allusion. The cosmic energy of the universe carries data of all kinds (cyberspace communications via cordless phones, radio, television, space-wide satellites, vehicles, and computer files) so transfer of data between some humans through and in cosmic space is probably occurring at currently. This is not a proven fact, but perhaps it is time to encourage earth scientists to focus on some ways to make the transfer between direct minds and the cosmos evolve, so the magic becomes useful in more work related situations. If the fund of knowledge in the cosmos is an accumulation of all the known data from the past, it might be monumental in its size and profundity, a worthy target for scientific minds to address.

Also, dreams indicate some connectivity. The mind, in sleep, is wide open to the input of cosmic energy and data. The data is communicated directly into subconscious awareness as a dream. As such, it has a relatively short life span in that state as it does not always, apparently, excite retentive memory cells. So it must be brought into conscious cognizance to be retained before it fades away. If one intends to make deliberate use of dream information, it must be quickly lodged in long-term memory to be of any future value.

The act of praying is a time-honored way to reach into cosmic space. The traditional approach to praying is to address a divine force that has been depicted as a humanoid figure, a god or reasonable facsimile. For most people, it is difficult to conceive dealing with an ethereal power for a blessing or special benefit. The orison is not unlike that which is expressed when looking for a responsive reaction through divine intervention, but the focus of pragmatic allusion would be quite different. With traditional prayer, one asks for direct assistance, a blessing, or some other boon from a recognized deity. This is contrary to the process when seeking a reaction from the cosmic resource.

If one hopes to receive a positive response from the atmosphere of the universe, it cannot be demanded or commanded. The process calls for an introspective viewpoint. The request must be phrased as a personal need and not as an order. The need must be infused as a challenge to one's own mind and be a request with intense emotional feeling. Then the pursuit can be abandoned as the communication of need will automatically be transferred to the cosmic energy field, and the action of responding will be fulfilled without conscious awareness.

This method of contacting cosmic energy may sound farfetched, but with a full belief in its effectiveness, the results may be surprising. Many who have failed to understand fully the magic that exists in cosmic communication may have experienced the sudden pop-up phenomena, and have succeeded or failed to effect a connection with the cosmic force.

The process of cosmic communications calls for the mental focus on two major cosmic laws. The first is that one who seeks to communicate with the field of knowledge in the universe (magna vault) must have deep, unequivocal belief in its possible existence as a viable force of the Universe. Without such indomitable belief, doubt would creep in, and the process would fail. The request also must be a desire, not an order. Prayers that request direct results may not be subject to cosmic awareness, so response may not be forthcoming. The speculation will intensify regarding cosmic communication and a breakthrough will occur in due time.

Personal Time and Space: Stopwatches and Mileposts on the Road of Life

Of some relevance to humans and perhaps to all forms of life on the face of planet Earth is the philosophical appreciation of how each one of us relates to the concepts of time and space.

There is some rationalizing among space scientists that, because they cannot be measured, cosmic time and space have no real relevance here as they differ in the way they are understood on Earth. This is illogical because both elements are highly important to the universe and also planet Earth.

Time, in cosmic terms, is eternal. It has one direction only, forward, and its function in the universe is to measure eons and eons of spatial change. It becomes a factor in the knowledge mankind when it is recognized as locating events in space. The totality of this epochal availability expands and contracts only to satisfy spatial demands that are made on it by components that occupy or have need to occupy the spatial milieu. Such components are the planets, stars, galaxies, comets, and other unknown types of possible space occupiers that may be filling voids in the cosmic atmosphere but are not yet formally identified.

A workable understanding of how most humans adapt their logical thinking to the presence of time in spatial terms calls for some philosophic perspective and awareness that reaches out beyond the standard day-to-day knowledge common to most.

Of importance is some recognition of what spatial time means to humans and how they appreciate but hardly ever consider it in terms of day-to-day existence. But it poses very interesting consequences at a deep and logical level.

Just think. At the moment when a new human life enters the free atmosphere after nine months in preparation, the offspring inherits a stretch of time, defined as destiny. This stretch is a set piece for everyone but may be cut short or elongated depending on two factors. These are parental genetic history for one and the cosmic law for all human longevity as the other. The tools of nature, all part of cosmic law, are in the supreme pleasure of procreation, survival instincts, and all the automatic functions of a living organism. Humans are destined to inhabit a very short period as Homo sapiens and cannot occupy beyond the limits they have inherited at birth. While birthing time can be predicted, dying time cannot. There are no rules for behavior in cosmic law, but the decisions to maximize or minimize the opportunity of living are left with humans to test their logic and intelligence. Perhaps some similar laws apply for all the occupants on the face of Earth, and while some elements may control their existence time, others may not. Time is precious to each living entity on Earth. If the human genre could visualize its destiny as a measure of finite time and view it as an inventory that they can add to or subtract from because of personal behaviors, a more serious and productive stretch of cosmic time could be attained for all living organisms.

Space, in cosmic terms, is infinite. Humans see it as a dome that has an atmosphere of electromagnetic waves and particles that are filled with suns, galaxies, planets, and black holes, among other elements. If galactic laws prevail, the universe may be globular, as are all other cosmic masses that have gravitational cores. So it may be logically supposed that the universe is similar, though we can see only a portion. If it is true, then full universal space is one monumental spheroid that could quite easily hold the Earth-known universe, as well as many others.

It is an established fact that the known universe is expanding. The new space therefore will be presenting content such as planets, stars, and galaxies. They are so far away that their presence cannot be easily detected at the present time, but creative developments on Earth may hurdle that spatial limitation one day. Perhaps what the Hubble telescope sees is the emerging side of the global universe that we have never seen. It could be a scientific challenge to see if the other edge of the universe is disappearing. If it is, the concept of the universe as a rotating globe would gain credibility. In any case, a revolving Universe would roll very, very slowly.

Because movement in the cosmos is evolutionary, changes take place over long, long periods of Earth time. But surely they do. By earthly evidence, however, evolutionary change can occur within the scope of one average human lifetime.

Unlike time, no cosmic laws limit the space that humans can occupy. Man-made laws can specify plots of land and volumes of air to limit or expand the size and time periods of one's occupation. For purposes of control, these are legal rules of ownership. However, in reality, no part of earthly space can be possessed. It can only be held for specific moments in time, but will always belong to planet Earth if ever man-made controls are terminated. The functions of rental, lease, and ownership are centered on the concept of accumulating and possessing wealth using the medium of planetary domain as the product of exchange.

Personal space, once claimed by documents of possession or just simple codes of existence, is worn about oneself like a platinum-coated cocoon. Its possessor learns to deem it precious, so it is not easily disowned. Attempts by strangers to invade that space without permission can be met with stern rebuffs and even violence. The human is his own body-space landlord.

Human personal space is incredibly mobile. It moves as a garment with each individual wherever that individual is free to roam. It will occupy space that others may own, but it asks no fee and pays no toll for the area it will use to contain the human that has reason to need usage time. Outside of that space possessed by legitimate occupiers, space on planet Earth is limited but completely free. All restrictions on its use and location are legally bound by those who think they have a right to exploit space on and around the planet Earth for their own mostly selfish interests.

Time and space are interdependent. One has no purpose without the other. When Homo sapiens have exhausted the ability of cosmic energy to warrant existing conditions for the ongoing survival of the genre, it, like all other exploited and overused planetary elements, will lose their right to exist in the grand design of the Universe. Universal rules would likely terminate it over time or cataclysmically.

Humans are not mature enough as yet to realize that perhaps the planet Earth is like a laboratory complete with specimens of which Homo sapiens is one. Human origins and evolution have been spectacular. From infancy, the world as we know it has charged all occupants to grow as a major experiment in the cause of human development. Needless to say, at this stage, we must profess that we have largely failed our mission so far. The human genre, as the major product of the experiment, has been given all the natural resources needed to survive and prosper as a predominant example of the type of creature that can care and share the bounties of our planet Earth. However, due to ignorance or lack of adequate intelligence, the traits of greed, overpowering egomania, the strive for fame and fortune, the quality of humankind has evolved from a concept of "one for all" to that of "all for one." Earth-based laws protect the owners who have amassed huge areas of space for themselves with no regards for others.

Readers Personal Space for Thoughts and Ideas

Synteresis: Ultimate Judging on the Dusty Roads of Destiny

In the studies of earthly humans, the constant puzzlement of how and why decisions are reached in matters of prime importance is at the top of the list. So many factors are involved that mental sourcing of justification for most psychological decisions is difficult to explain because they ask for answers from the deepest sense of human emotions.

What power in the human mind can juggle the pros and cons, the good versus bad, and the positive and negative likely consequences of the final decisions reached? In trying to arrive at some logical conclusions, a visible structure of decision-making can be helpful. An example of such a structure can be found in the doorways of the well-known megalithic ruin known as Stonehenge, located near a town called Amesbury in the county of Wiltshire, England. In its stone circle, there are several door-shaped openings with two separate columnar pillars united at the top by a lintel. In general views, they could be defined as an upturned, square letter "U." Selecting one of the larger shapes as an example, the structure of decision-making can be logically organized. These are the portals to the corridor of learning.

The necessary elements are simple. One of the pillars can be defined as the positive, humane feelings and attitudes in the human psyche. As all positive characteristics in our known society must be matched with the negative ones, the second pillar will define the negative feelings, attitudes, postures, and failings of the people in the structure, everyone.

Across the top, joining both pillars, is the lintel that contains the human values that support decision-making, and above that must then be a supreme, final, irrevocable decider. This is the one where the ultimate value systems become the synteresis of judgment. With these factors, it is possible to create a logical structure and functional validity as a design for human decision-making. The diagram will illustrate the concept in a viewable form and sets the pillars as the values that are involved in reaching presumed or correct decisions. Communication between the pillars and thence to the lintel occurs at the speed of light as neurons that are electromagnetic and constantly in motion control it.

Synteresis combines the influence of virtues, failings, and the impact of pragmatic allusions on the confidence required to make proper decisions and/or sustain conflictive communications.

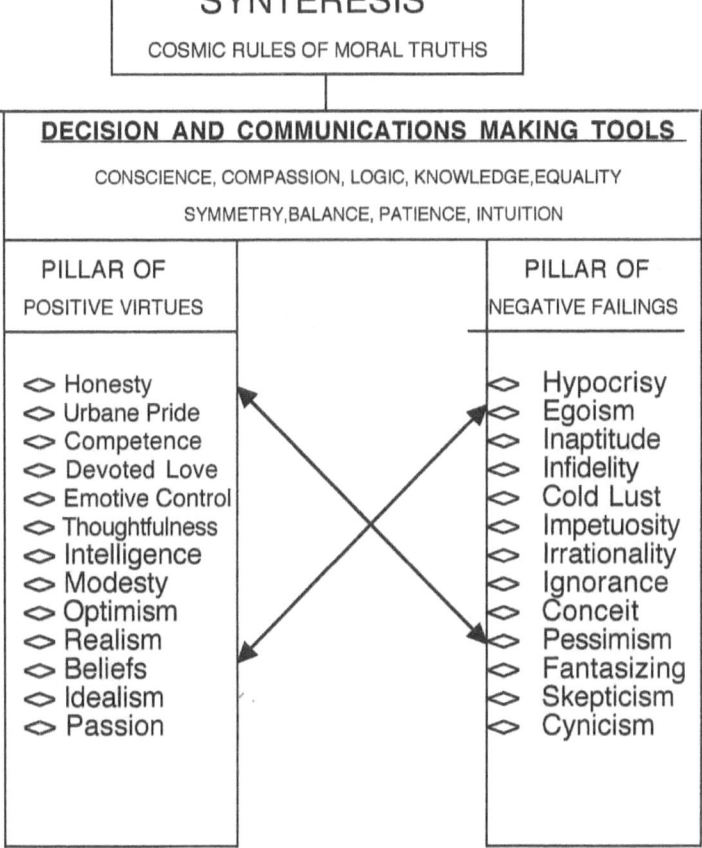

The above two pillars are representative of the positive and negative character traits of average human beings. Because humans are a gregarious and adventurous lot, there are many more virtues and failings in the genre beyond those above.

Communicative intercourse and decision-making call for interplay between the two pillars with lightning speed. Most common decisions and chatter are resolved within one or other of the pillars as the attitudes establish themselves and confidence evolves. If they cannot be finalized by interplay, then the problems move up into the supporting sector. Here a more confrontational exchange of views is likely, and if the compromises fail to bring resolution to the issue or issues, the last chance in the situation will be to run the conflicts up the decision-making line to the top of the arch.

The last resort there evolves in synteresis, where a judgment must then be handed down. Final and irrevocable. The process may be long but effective.

Synterisis brings decisions and all communications finally into contact with the energy of cosmic electromagnetic logic. While it would allow both parties in an event to use normal attitudes (both virtues and failings) for judgmental purposes, synteresis (the belief that common logic prevails) adds the third dimension with cyber symbiosis to elevate the human capabilities to a higher level of intelligence, awareness, and pure logic.

The process of decision-making on most occasions does proceed at the primary levels of communication. However, in the present day and age, it seems that many events demanding precise direction invite a broader range of opinions and hence embodies the full range of emotional attitudes. Because the pace of enterprise and development is escalating, some study of the psychological elements involved in key matters can be time-saving and perhaps most effective. The syntereal process can prove two positive outcomes: faster action decisions for one and a focused, longer-term professional process through training and experience for the other.

So, as long as the accumulation of wealth is to remain the basic mantra of earthly humankind, increased capabilities to move wealth-producing events forward will be evolutionary prerequisites. When the present capabilities of the human mind are stretched into new and untested territory, evolutionary forces could be imposed to ensure that progress will prevail.

Readers Personal Space for Thoughts and Ideas

Character: Proof of Soul and Spirit

Character is a measurement of what all humans evolve to from birth to death. It defines how ingrained are the basic qualities that are burned into the human psyche by genetics, education, parental guidance, rare experiences, diseases, and the rough and tumble of day-to-day survival.

The inborn seed of humankind from which the character of every individual person is created is the belief that cosmic energy sets the stage upon which tots and teenagers, not to exclude mature adults, perform their lifetime mundane and creative acts of existence. The very early stages of a developing human creature are inbred with cosmic logic based on the credo that normal humans are all born with a native sense of right and wrong, good and evil, acceptance and denial, and pain and pleasure. At the early stages of existence, the human has the way of knowing only the whats of situations and events, but not yet the whys. Then they gradually learn the difference between the pros and cons of life's experiences as maturity begins to imprint its special influences. At this point, a character begins to evolve and will be etched into the nature of the individual at every level and moment of their development

So, how does one define the nitty-gritty of character? Onlookers can see and experience it as an indefinable aura that surrounds its possessor and is called presence, perhaps even personality. This quality radiates confidence and promises security. It presents attractiveness even though the reality may be less than the usual standards on which beauty is measured. It virtually guarantees integrity and flings open the door to compatibility. In any language, a person with character draws the power of popularity from a person or audience. This power is intrinsic in the atmosphere and creates the feelings of amiability.

Character is the method displayed to one and all everywhere of the spatial place of humans in earthly time and space. Without a visible handle, it is difficult for many to mentally visualize their presence in the monumental area of the electromagnetic universe. Our individual importance to the cosmic atmosphere is moot. It is not measurable. However, the collective importance of all humans does impose a standard of behavior that underlies the basic awareness of all individual persons. This aura is adhesive, and it sticks to one with character like iron filings to a magnet.

Every human has the makings of a real or pseudo image of character. It cannot be bought or borrowed, but the femmes of fate can hang an evil or good mantle of deportment on the frame of anyone whose choice of overt living is rife with greed, cruelty, selfishness, bigotry, and animosity. Conversely, with compassion, affection, compromise, empathy, and tolerance.

By comparison, personality can be defined as the active ramification of character, but it may lack the depth of values that are the bulwarks of the broader rules of behavior. While personality is usually most effective in person-to-person communications and can coat the real human in the process, character resides in the soul and never leaves the mind and body whatever the occasion may demand. Personality is not required in the quiet solitudes of innate thought, but character never departs from the living mind of humankind.

Ahhh, deep within their sentient souls
Reside all human forms.
In moments, private and complex withal,
They muse on living norms.
To give and take, the dancing fates decide
What should be best for all.
Create some lurid, dark events.
Plus others to enthrall.
These states of grace, they must embrace
Their spirit to employ
In searching cosmic time and space
For humans to enjoy.
To wear these cloaks and gain respect,
That's well dressed in the mind.
They must seek robes of guileless pride
And leave vain togs behind.

The true hue of what lies beneath
That cloak of human skin
Lets character come shining through,
To limn the truth within.

Readers Personal Space for Thoughts and Ideas

Faith, Hope, and Charity:
Redeeming Virtues in a Tumble-Down World

The saving graces on a planet where freedom is promoted, rights are protected, human guilt is suborned by baser values, and exploitation is rampant, those graces that survive are those that are deeply inscribed on the genetic blueprints of living human beings.

These historic traits serve well their roles of preserving basic honesty, lauding diligent effort, revering true humility, extolling compassion, and proselytizing true love with all its cloaks of virtue and vicissitude. Like morning mists o'er the meadows of life, faith continues to support the better qualities of human existence, hope still tests the vision in seeing better things on the way ahead, and charity explores the symbiosis that exists between human souls communing in their garments of diverse design and color. The soul of humankind is the very essence of all and fulminates against the challenges to the true virtues that make humans what they really are.

The articles of faith with humans are values that are owned at birth but are questioned as wisdom matures and experiences teach. The fundamental fathers of the concepts of faith were early shamans, who, through the use of sacrificial rituals and fetishism, provoked the fear of sorcery and demonism. Awe, embedded in the mystery of ignorant beliefs, bore the heavy lifting of faith into the future.

Recognizing the power that exists in faith to affect and control large masses of humanity begat the known concepts and interim principles of current forms of faith that govern human destinies. These entranced the spirit of humankind by convincing, early on, that gods of many stripes existed and controlled their fate during earthly lifetimes and perchance subsequently as well. The evolution of intelligence has latterly created suspicion and incredibility among more learned minds. They continue to reach out beyond the concepts that gods exist on Mount Olympus, eating ambrosia and imbibing nectar while playing chess with unmanageable human beings imbued with peculiar peccadilloes.

In this out-of-the-box maturity, the theories of a polytheistic existence began to fade into obscurity, while the neo-philosophy of one god over all took root. There are many religions, but one universal god whose existence has been conceived as belonging to many genres of followers. Thus Christianity, Islam, Buddhism, Hinduism, and Judaism have become the mantras of this supreme god with different names, and with fringe beliefs stringing along with their minor congregations scattered and somewhat inert.

Definitive manuscripts, with historical credentials, such as the Holy Bible, the Koran, and the Torah serve the needs of the various congregations and their dedication to convert humankind. Those whose belief transcends all examples of visible reality are deeply committed and hence consider themselves divine. With a sterling state of mind, they brook no challenges to their definitions of belief. While many others may decry such dedication, they cannot deny believers of the right to their beliefs.

Faith and hope are kissing cousins living in the houses of belief. Faith sustains the thought that something real exists within the concepts of dogma, while hope is a deep-seated desire to find filaments of reality in the families of faith against rather dubious odds. With hard-bitten logic, it seems that the human soul would wither and decay without the manna of hope to allay all doubts. Hope has many costumes and almost universal congregations. It is the motivator of creative energy and can sustain a target objective through many frustrating reverses. Without hope, a deadly ennui might engulf the world, and humankind could sink into desperate oblivion.

Charity has both real and intrinsic virtues. A sense of dedication to both worthy and selfish causes can prick a sense of charity to emerge. In philosophic terms, it defines the term of "do unto others what you would have others do unto you." This limns the idea that emotional equality can and does exist in a human world and to do good for others has its own rewards.

Because it has its roots in human ideology, charity has its negative aspects up-front in human exploitation of both faith and hope. The calls for charitable mercenary donations are innumerable. This would suggest that exploitation of the goodwill sense inherent in the souls of mankind can leave doorways open for ill-gained opportunities. The power of money has no conscience when it comes to greedy aims and ambitions and can destroy the integrity that is the very core of charity. So, like many schemes that target human compassion and goodwill for personal and often prurient reasons, the true virtue of charity gets lost in floods of uncharitable coins. With the spirit eroded, the real meaning of charity dissolves.

Faith, hope, and charity are all attitudes invoked by humans on this planet Earth. They are three candles in the wispy wind of feelings that are multifarious and inconstant in multifaceted oceans of humanity. Just how strong these virtues may be will probably be tested in the next phases of evolution that is constant and eternal. Perhaps the short life spans of humans are a boon to be enjoyed as a cosmic privilege. Death, in all cases, levels the boulder-ridden landscape for everyone. It is an unfortunate reality that humans then cannot enjoy the level horizon of a perfect, peaceful landscape. At least under current ideology.

Readers Personal Space for Thoughts and Ideas

Cosmic Space and Planet Earth

In making any relative comparison between the space volumes of planet Earth and the volume of the Universe, it begets the vision of a grain of sand, a mere particle in the middle of the Sahara Desert or a drop of water in any ocean of this world. Like the atom, small particles contain even smaller parts.

However, the influence of cosmic energy over planet Earth cannot be dismissed. Planet Earth is in the Milky Way, and that galaxy is a major structure in the universe that humans can observe. Looking at the grandeur of the Universe with earthly eyes and speculating on the meaning of it all calls for huge packets of creative imagination to even come close to understanding what one is seeing and logically explaining what one thinks might be going on out there. This speculation need not be cloistered only in the labs and minds of scientific communities, but can also be present in the thoughts and attitudes of lay thinkers, dreamers, and those with the courage to visualize events beyond the box of a mundane imagination. The insignificance of a human body in cosmic space tends to reduce human confidence to the generality of an invisible idea.

To rise above those insecurities takes courage and dedication far above a normal challenge.

The Concept of the Universe

Like with all concepts of events using Earth-formed logic, they have to have a beginning and an end. This theory has led to scientific communities accepting an answer to the conundrum of the Universe as having its beginnings in an event called the Big Bang, but no scientifically acceptable point to closing down the cosmos. There are many unanswered questions with this theory as to the beginnings of a universe from a gigantic explosion of elements. This theory skips over the fact that no singularity can erupt from thin air, so credentials for the big bang are believed to have been founded in religious dogma that supports the idea of one omnipotent godhead, who, using a magic wand, created the Universe and all contained within its unlimited space. No realist can believe that underlying forces of a more scientific nature could not have been involved. So be it. Everyone is entitled to opinions but must bear the burden of proof if an idea has no traction in the mind of human intelligence. Without this proof, the door is open to other concepts of the universe. Some new ideas are every human's right to propose in the absence of credible, logical answers to the purposes, mechanics, and logic of this mighty cosmic occupant in open space that might be called the void.

Defined Views Regarding the Cosmos

1. Space, in the Void wherein the cosmos exists, is infinite. This means there are no limits to the area that celestial bodies of atmospheric gases or material mass can occupy. This implies that the void will always supply the needs for anything that requires room in the universe.

2. Time in the Void that influences cosmic events is eternal. This sets up the conjecture that the past no longer has relevance in present time as a measuring device other than to record events. Events will have existed to give body to the concept of evolution and its impact on the shape and size of all future transitional developments.

3. Evolution is the process by which changes in the Universe are instigated. Evolution moves the present into the future, forward into new space that is made available, if need be, from the inventory of the Void.

5. Gravity is the power that attracts all cosmic contents from and to any point in the universe and even in the void. Gravity force exists everywhere. It is present in the mass of bodies and the electromagnetic particles that inhabit the universe and perhaps the void. It is interminable and defies the intelligence of human minds to parse its composition, creation, and maintenance.

6. Dark energy provides the motive force to events that evolve in space. It has a primary role in creating, maintaining, and making an inventory of the powers of change in the universe. It exists wherever change is taking place.

7. Black holes might be cited as the garbage bins of the universe. They are almost invsible to humankind but have incredible power to attract and subsume any mass with gravity less powerful than that of the black hole. Star/suns that have used up their effective energy become white stars, devolve into black holes themselves, or fall into the core of an existing black hole with the superior attraction of gravity. The disposition of material that has been attracted is not entirely clear, but it is thought, because of the intense heat existent there, that material will change its form into dark energy and be returned to the inventory of same. Once thought of as rare, black holes are now located with superior technology and found to be profuse through the cosmos.

8. Motion in the universe was once thought to be rare and insignificant. In fact, the universe is a virtual mix master of change. Because it occurs so far in space from planet Earth and Earth time is measured in eons, the movements of change have been misjudged. No longer so.

9. Atmosphere is the invisible coat that cloaks the universe with a blanket of electromagnetic energy that permeates the air but also every sentient and inert material contained within. It is moot to think that this energy also infuses the space that makes up the void. Logically, only elements that consist of mass material or are of a gaseous or atomic composition can be imbued with EM energy. The wireless instruments of communication, entertainment, and technology provides evidence of this energy. They transfer data at the speed of light throughout the cosmos and is used on Earth. Whatever other components exist in the atmosphere are known to the scientific community, but it is accepted that there may be others that have yet to be identified.

10. By using Earth-based vision and ideology, the Universe is viewed from the surface of the globe as a flat or arched endless panorama like a map. All planets, stars (as suns and inert bodies), and other masses within this spatial volume are relatively immobile, although there is incessant motion taking place over periods of Earth time with no specific magnitude. The constant expansion of the universe is an event of some concern but is viewed only with the help of modern technological instruments (Hubble space telescope) that confirm this expansive movement.

11. The solar system is seen as a huge sheet of stars/ suns set in galaxies and super galaxies throughout the cosmic space. The suns may have a host of planets circling them in different orbits or be in process of developing or declining in importance as their active lives begin to deteriorate. The planets themselves, if they have a magnitude that is adequate to have a critical gravitational core that can attract and hold satellite bodies in orbit, will develop specific atmospheres on such planets, suited to their reasons for existence. The location of the planet Earth, by design or happenstance, has been ideally sited to promote the creation and survival of Homo sapiens and all the supporting flora and fauna critical to such life on Earth.

New Thinking on the Universe

The holes in logic as it relates to the origin of the Universe and the subsequent rationales that have been engendered have created some real pressure for other alternatives. What is to follow is one trail of ideation that has produced some logical arguments for change in the philosophical and practical ideology that currently obsesses the thinking of the cosmologists, mathematicians, and spatial scientists.

A Different View of Space Logistics

1. All forms of matter in the cosmos that embody gravitational attraction find their natural shape as a globe. From the Earth itself, the sun, the moon, the planets, and even cellular structures such as atoms find the global form the ideal solution to the challenges of gravity. If this is seen to be a cosmic reality, then the shape of the Universe and the spatial void must take the same configuration. If the human mind can comprehend the size of the Universe, then it should be able to see it as a huge circular ball within the infinite void of outer space. If it is a global system, then it must be rotating around a gravitational center of immense powers of attraction that is holding the entire structure of the cosmos and its contents together. If it is logical to support the theory of a spheroid universe, it may suggest a logical solution to the mystery of its expanding space. It may take great imagination, but if it is a sphere, the universe is gradually turning on its gravitational core, and a new side of the ball-like structure is coming into view. Albeit, very slowly said, to be some 2 to 3 percent per annum. However, in view of the fact that, in Earth terms, time in space is measured in eons, single generations of humans cannot note a small change. The fact that an unseen side of the global universe may be appearing is a basic factor

in establishing that the universe must be, by logical extrapolation, a spheroid structure.

2. With that thought in mind, to expand the imagination, one might see that the Universe itself is contained in a Magnaverse that contains many universes within its globular confines. If it can be accepted that the space in the void is infinite, all bells can ring with the truth behind the general shape and size of the universal reality.

3. Black energy is the positive force of change in the universe. It attends to all the requirements that make evolution a progressive tool for advancement into the future. It is produced by the stars, whose fiery hearts convert hydrogen to helium and sundry other forms of chemical change. In the process, black energy emerges and is transferred to the atmosphere of planets as well as to the planets themselves. The energy that infuses planets like Earth is all-pervasive and found in all forms of matter and environmental elements. Humans are in constant exposure to one form of black energy that is labeled "electromagnetic force." All positive forces must have complementary negative energy components. These forces serve as controlling facets to ensure that the conditions of balance and symmetry can be ensured. The positive energy must be supreme, or any system will collapse because of the negative effects.

4. Evolution is the omnipotent force that compels constant change in the universe and all its component parts. The power of this cosmic tool is felt on planet Earth as the indomitable presence of entropy and atrophy in all forms of matter that cease to develop and maintain the basics of survival. The life cycles of all things are the blueprints of evolution.

5. Reproduction is the platform of survival that is part of the evolutionary process. The will and power to create new entities from the shadows of the forbearers is the role of reproduction in the cosmos. Humans may be the only genre to use technology to bypass the pressure to reproduce with chemical and instrumental devices. The long-term consequences of abortion and contraception are still fully to be realized, but because it is a cosmic law, dire changes in this area may be anticipated. The conflict for humans lies in the pleasure of sexual events and the consequences if parents are unable to rear the offspring in a proper manner. In the universe, the production and maintenance of the assets of the system are in constant change. Stars are forming and turning into black holes to vanish when their energy is expended. In any alternate configuration, they may become white stellar forms. In the core of other cosmic actions, the acts of reestablishing the functions that have strayed from symmetric balance are under

constant surveillance and adjustment, but to Earth-bound eyes, they are invisible.

6. Symmetry and balance in the cosmic system is an important method of ensuring a smooth continuity to all universal component activities. The force to bring a return to normality of balance and symmetry from aberrations is seen as a mandatory rule. And because of the long periods of Earth time that this can involve, the spatial changes are not generally perceived nor monitored within the time frames of human life spans. Cosmic history becomes invaluable.

7. Survival is an instinct that seems to pervade the universe. The instinct to defy the finality of death is deep-rooted in the human species, and this begets the thought that it retains its power over the will of temporal and religious ideology in all segments of the cosmos. While it ensures the continuity of all segments within the universal system of operations, it may not have the same mandates in places like Earth, where independent thought has become a powerful counter force.

8. The singular purpose of the universe has never been seen to occupy the time and effort of the space scientists because it would be a speculation and hence a rather inefficient use of time. With some bending of the neurons in the mind, this may have been visualized

as an exercise in justifying earlier conclusions reached about the universe itself, the main one being god as the magician. From a rather distant viewpoint in cosmology, the only real purpose for this hugely complicated and omnipotent structure is seen as the master progenitor of all things in the cosmic void. In some ways, this might appear as God the Almighty. However, it departs the religious ramparts to become the creator, maintained monitor, and ultimate historian of all possible living elements within the responsibility of universal objectives. Under this ruling of cosmic purpose, the Universal system has existed for far longer than the big bang event. Because there are no limits on time or space in the universe, the need of a cosmic start or finish time is irrelevant.

9. Communications within the universe are thought to be through a system not unlike the neuron network of the human body. Information travels at the speed of light and hence appears to be instantaneous to human eyes. Communications between the universe and humans has not yet developed into an overt, directly understandable language. However, since the atmosphere of the cosmos and the subconscious mind of humans are united by the atmospheric electromagnetic energy that permeates both equally effectively, the connection must exist but as yet is indefinable. Contact has been claimed

by many who extol their experiences with so-called magical events that have no other rational explanation. Reaching into the bowels of the cosmic electromagnetic communication vault is no easy task for humans with no experience in finding the keys to electronic histrionics. One fact emerges. To get response to a specific need from the universe is to be a deferential personality. Fix the issue firmly in one's own subconscious. Express the need for a response to oneself. Then set the issue to one side. If a response is valid, it will appear from the cosmic source in the sender's conscious mind PDQ. Properly executed, the process works!

In closing up this cosmic cybertrek window, it has been a very commanding experience to have found extension in the mind sufficient to undertake this exploration into space. One day when human minds mature even more, this trek will be a common-day breeze. To live long enough to experience that event would be a gift. Ciao.

Readers Personal Space for Thoughts and Ideas

About the Author

William A. Salo was born on August 18, 1921, and raised in Canada. He was educated at University of Toronto and graduated with a B.Comm. degree in 1949. He served in RCAF during WWII as a pilot from 1942 to 1945. He married his wife, Lois, in 1952 and they have been married for over sixty years. Their three children were born in 1955, 1958, and 1964. He held a career with two large corporations with senior marketing responsibilities. With a partner, he owned a large Canadian advertising agency located in Toronto, Canada. Thereafter he had full ownership of a management consultancy with a worldwide clientele. He fully retired in 1978 to a life of writing and many other mentally challenging pursuits.

www.ingramcontent.com/pod-product-compliance
Lightning Source LLC
Chambersburg PA
CBHW030300290526
45785CB00001B/154